Per Axel Rydberg:
A Biography, Bibliography and List of His Taxa

Arnold Tiehm and
Frans A. Stafleu

NYBG

The New York Botanical Garden
Bronx, New York 10458, U.S.A.
Issued: 13 March 1990

MEMOIRS OF THE NEW YORK BOTANICAL GARDEN VOLUME 58

COVER: *Hymenoxys grandiflora* (Pursh) Parker (*Rydbergia grandiflora* (T. & G.) Greene). Rydbergia. Illustration by Charles F. Yocom, in **Rocky Mountain Flora** by William A. Weber, published by the University Press of Colorado. Redrawn by Bobbi Angell.

Published by
The New York Botanical Garden
Bronx, New York 10458

International Standard Serial Number 0071-5794

Library of Congress Cataloging-in-Publication Data

Tiehm, Arnold.
Per Axel Rydberg : a biography, bibliography, and list of his taxa / Arnold Tiehm and Frans A. Stafleu.
p. cm.—(Memoirs of the New York Botanical Garden ; v. 58)
ISBN 0-89327-351-1
1. Rydberg, Per Axel, 1860–1931. 2. Botanists—United States—Biography. 3. Botany—North America—History. 4. Botany—Bibliography. I. Stafleu, Frans Antonie. II. Title. III. Series.
QK1.N525 vol. 58
[QK31.R9]
581 s—dc20
[581′.092]
[B] 89-29406
CIP

International Standard Book Number 0-89327-351-1

Frontispiece. Per Axel Rydberg (1860–1931). Undated photograph, courtesy of Hunt Institute for Botanical Documentation.

MARCH 1990 MEMOIRS OF THE NEW YORK BOTANICAL GARDEN **58:** 1–75

Per Axel Rydberg: A Biography, Bibliography, and List of His Taxa

ARNOLD TIEHM[1] AND FRANS A. STAFLEU[2]

Contents

Abstract ... 1
Introduction (A. TIEHM) ... 1
Per Axel Rydberg (1860–1931): A biography (F. A. STAFLEU & A. TIEHM) ... 2
A Rydberg Eponymy (A. TIEHM) ... 19
Bibliography of P. A. Rydberg (A. TIEHM) ... 20
New Taxa Described by Per Axel Rydberg (A. TIEHM) ... 25

Abstract

Tiehm, Arnold (New York Botanical Garden, Bronx, NY, 10458-5126, U.S.A.—Current address: P.O. Box 11278, Reno, NV 89510–2278, U.S.A.) and Frans A. Stafleu (Institute of Systematic Botany, Heidelberglaan 2, P.O. Box 80.102, 3508 TC Utrecht, Netherlands). Per Axel Rydberg: A biography, bibliography and list of his taxa. Mem. New York Bot. Gard. **58:** 1–75. 1990. A biography of Per Axel Rydberg (1860–1931), including a 215 article bibliography, is presented. Two generic eponyms and 34 specific and infraspecific taxa named for Rydberg are listed. An annotated account of the 1976 new taxa of plants described by Rydberg is provided.

Introduction

In November of 1983 the New York Botanical Garden was preparing an index to the microfiche of its vascular plant type collection. Patricia K. Holmgren assigned me a small task of checking a Rydberg type, casually commenting that someone should "do Rydberg's life and publications and if I had the time I would do it myself." A few days later, while visiting the Gray Herbarium, I mentioned the idea of a Rydberg project to Reed C. Rollins and Richard S. Cowan, both of whom enthusiastically endorsed the idea. Upon returning to New York I consulted Holmgren about the possibilities of such a project and received her wholehearted approval.

As this project progressed, both Holmgren and Rupert C. Barneby impressed upon me the point that anyone ever encountering a problem with one of Rydberg's types, publications, or publication dates should be able to solve it by using my work. I strove to this end but I am not so naive as to think my work is perfect. There are still problems with a few types, as can be seen in the following list of taxa described by Rydberg. However, I have presented the best information that I can assemble at this time.

Recent books on Rydberg's contemporaries, Aven Nelson and Marcus Eugene Jones, have shed light on that era of North American botany and on his vintage of botanists. These were colorful and at times controversial individuals. Nelson had many students, served as President of the University of Wyoming, and was involved in some of the controversies concerning the American and International codes of botanical nomenclature and hence conflicts between the New York Botanical Garden and the Gray Her-

[1] New York Botanical Garden, Bronx, New York 10458. Present address: P.O. Box 12278, Reno, Nevada 89510-2278, U.S.A.

[2] Institute of Systematic Botany, Heidelberglaan 2, P.O. Box 80.102, 3508 TC Utrecht, The Netherlands.

barium of Harvard University. Jones was an outspoken man who was often retained as an expert botanical or geological court witness and as such was one of the foremost environmentalists of his time. In his later years, he became quite caustic toward his contemporaries and through the pages of his own journal, *Contributions to Western Botany,* he printed many a scathing article (often an obituary). Despite this, Jones should be remembered as an intrepid collector and botanist who appreciated the influence of environment on the evolution and distribution of plants.

Rydberg was neither a Nelson nor a Jones. He was a gentleman of the old school, who never wrote defamatory articles about his peers and stood apart from political and, until cornered at Ithaca, methodological controversy. His private papers were destroyed by his family after his death, but there is no reason to conjecture that they would have revealed a character other than the shy, upright votary of science that is recalled in brief memorials by his colleagues at the New York Botanical Garden. His monument is his work, which testifies to a strong physical constitution and to a lifetime's devotion to botany.

In a sagacious and detailed pre-publication review of this paper, a reviewer suggested that the biography could be materially enhanced by a review of the present status of all taxa proposed by Rydberg in the course of his career. Such an evaluation would facilitate comparison of Rydberg's work with that of Edward Lee Greene, Jones, and Nelson, for each of whom an itemized account is in the published record. The proposal has obvious merit, but its realization is far from easy. Greene worked primarily in extratropical western North America; Jones, if we overlook his ill-advised incursion, late in life, into the Mexican flora, worked almost exclusively in western United States; and Nelson almost exclusively in the Rocky Mountain and Intermountain States. Correlations of modern floras and monographs have yielded verdicts, in Nelson's case melancholy, in that of Jones mainly favorable, in that of Greene ambiguous, on the durability of their taxonomic opinions, and a comparable assessment could now be made without difficulty of Rydberg's record in extratropical latitudes. But Rydberg, in his contributions to *North American Flora,* covered a much wider field, ranging from Alaska to Panama and the West Indies. For the region south of the Mexican border we have only scattered and incomplete floristic literature and a very imperfect monographic literature. Until the several tropical North American floras currently underway are published and modern revisions of, for example, Mexican and Central American Asteraceae and Fabaceae become available, no full assessment of Rydberg's revisionary work can be made.

Acknowledgments

I am grateful to the following people who have unselfishly aided me with this project, which could not have been completed without their aid: Frank Almeda, Henrik Balslev, Hollis G. Bedell, Margaret R. Bolick, Mona Bourell, Jane Brennan, Barbara A. Callahan, Steven P. Churchill, Lynn G. Clark, Arthur Cronquist, Roy Curtiss, Thomas J. Delendick, Barbara J. Ertter, Susan Fraser, Amy Jean Gilmartin, Sarah Goldstein, James W. Grimes, Elizabeth Hall, Lawrence R. Heckard, Barbara J. Hellenthal, Martha Hill, Noel H. Holmgren, Alice Q. Howard, Anita L. Karg, Matthew T. Lavin, Lothian Lynas, Harold N. Moldenke, Nancy R. Morin, Gerald B. Ownbey, Alma Plantenberg, Deborah A. Qualls, Peter H. Raven, Kathryn R. Rollins, Reed C. Rollins, Alfred E. Schuyler, George Small, Jeffrey L. Strachan, Mary Thomas, Theodore van Bruggen, Dieter H. Wilken, Roger L. Williams, and Steven J. Wolf; to William A. Weber and an anonymous reviewer for their critical reviews and helpful suggestions; to the curators of the following herbaria who allowed access to their collections: CAS, DS, GH, JEPS, NDG, NY, RENO, RM, UC, US, YU; and for support from National Science Foundation grants BSR-8318256 and BSR-8618970 to the New York Botanical Garden while searching for Rydberg types. After the project was otherwise completed, Prof. Ove Almborn of the Botanical Museum of Lund, Sweden, kindly assisted in the translation from the Swedish of Rydberg's first publication and provided a duplicate type for NY. Of special note is the encouragement and advice I received from Rupert C. Barneby and Patricia K. Holmgren throughout the duration of this project.

Per Axel Rydberg (1860–1931): A Biography

Rydberg (frontispiece) was one of the most productive scientists at the New York Botanical

Garden (NYBG) in its first century, a botanist with a comprehensive knowledge of the flora of the Rocky Mountain region and of the Great Plains of North America; but he remained somewhat in the shadows, both in his lifetime and after his death. He was the subject of severe criticism, directed not at his personality, but at his scientific concepts. Rydberg, in short, suffered from dissent and neglect, notwithstanding his gentle character and behavior. As he stood apart from the professional disputes of the period, it is not easy to understand why his excellent qualities never received the publicity aroused by his—in some ways indeed unusual, and often disputed—professional ideas. He was, according to his colleague John Hendley Barnhart, who wrote the main obituary, "faithful to the service of his science and his God" (Barnhart, 1931). Apart from Barnhart's short obituary, Rydberg's personality and work have never been the subject of a detailed biography. This, however, is not exceptional. The only "biographical memoir" of any of Nathaniel Lord Britton's collaborators was that of Marshall Avery Howe, an honor derived automatically from his membership in the National Academy of Science (Setchell, 1938). Britton himself, also a member of this august body, had his biography published in the same way (Merrill, 1938). No other staff member of the Britton period found his "sympathetic biographer" (Williams, 1984). In fact, neither Britton nor the New York Botanical Garden, let alone Rydberg, has so far been the subject of a special study.

Per Axel Rydberg was born on 6 July 1860, within eight months of the publication of Darwin's *Origin of Species,* nearly eighteen months after the birth of Britton, the man who would dominate his life, and sixteen months after Aven Nelson, his principal competitor in western United States botany, His parents, farmers in Odh, Molla Parish, Västergötland, Sweden, moved to Lofvene near Lidköping a few years after the birth of their son. At thirteen their son was sent to the preparatory school for the Royal Gymnasium at Skara (founded in 1641). Here Rydberg made a herbarium and graduated in 1881. In this same year he submitted a half page botanical note to *Botaniska Notiser,* at Lund, in which he described a new variety (var. *acaulis*) of *Cardamine pratensis* (Rydberg, 1882). This was the modest but early beginning by a man who was to publish over 7000 pages of botanical studies.

Rydberg emigrated to the United States in 1882 and found a job in the iron mines of Michigan. Nothing is known of that period, except that in 1884 he met with a serious accident from which, according to Barnhart, he acquired a lameness in one leg that became more noticeable in later years, especially when he was tired. He had been trained as a mining engineer, but his Gymnasium education had also prepared him for less remunerative but perhaps more stimulating "intellectual pursuits." He became a teacher at Luther Academy (later known as Luther College) at Wahoo, Nebraska, a denominational college befitting his Swedish-religious upbringing. Rydberg was put in charge of the classes in mathematics. Seven years later he sent his first letter to Britton, mentioning his accident (Rydberg to Britton, 13 Oct 1891, NYBG archives): "Seven years ago I made a fall of 40 feet and strained my back which made me unfit for botanizing the first years of my stay in Nebraska. . . ."

He remained at Luther Academy, with some interruptions, until 1893. While at Wahoo he registered at the University of Nebraska, Lincoln, where he studied botany under Charles Edwin Bessey, about whom John Merle Coulter wrote "Professor Bessey was a most stimulating teacher, and perhaps no American botanist has left his mark upon so many students" (Coulter, 1915).

Most of Rydberg's work for his B.S., obtained in 1891, was done in the school year 1890–1891. From 1884 he had been unfit for botanizing, but in the summer of 1889 he went to Upper Michigan to collect. In his introductory letter to Britton (1891) he mentioned that "only the two last summers I have made any collections to speak of in Nebraska." During the summer and part of the fall he was employed by the United States Department of Agriculture (USDA) as a collector in western Nebraska. Of his duplicate collections three went to Washington, D.C., the rest stayed at Lincoln, but he sent duplicates from his own herbarium to New York. Many of his discoveries which were new state records were published by Bessey in a second appendix to Webber's catalogue of the *Flora of Nebraska* (Bessey, 1892). An account of Rydberg's 1891 collecting trip, which has survived in carbon form, tells of his refusal to botanize on Sundays, for purely religious reasons (Kiener, 1951). In this matter of principle, if in no other way, he resembled Marcus E. Jones.

In a letter to Britton dated December 1890 (Bessey to Britton, NYBG archives), Bessey announced the gift of a package of plants and his intention to send Rydberg's annotations. "He is now spending a year with us, and is devoting considerable time to botany. You will find his notes quite instructive." The package contained specimens of seven species of *Psoralea* accompanied by critical notes by Rydberg. "You must make some allowance for Mr. Rydberg's English, as he has been in this country but a short time, and has not yet completely mastered the English language. I told you of Mr. Rydberg that he has been studying with me, and is, I think doing very good work."

After the field work of 1891 Rydberg returned to Luther Academy to teach natural sciences and mathematics. In the summer of 1892 he undertook another collecting trip for the USDA, this time to the Black Hills of South Dakota. In the summer of 1893 he collected for the same agency in western Nebraska. He was also engaged in the Botanical Survey of Nebraska with five other students, among them Roscoe Pound and Frederic Edward Clements. This activity resulted in a *Flora of Nebraska* project of which only parts 1, 2, and 21 out of a projected 25 were published. Part 21 was published December 1895 and contained the Rosales by Rydberg (Rydberg, 1895a). The same year Rydberg (1895b) received his M.S. from the University of Nebraska and subsequently published his *Flora of the Sand Hills of Nebraska* (Rydberg, 1895c), based on the thesis. In both publications Rydberg had to consider the genus *Astragalus,* which was to become one of his specialties in later years. His treatment of the Rosales was reviewed by John Kunkel Small, who would become a colleague for many years at the Garden. Small's review was very positive: he praised the original plates and mentioned especially the fact that all species of *Astragalus* were figured (Small, 1896).

In the summer of 1895 Rydberg collected in Montana, with Cornelius Lott Shear, again for the USDA. The particular charge was to collect grasses and other forage plants, but he also dealt with the general flora.

From the fall of 1895 until May 1899 Rydberg was, with some interruptions, professor of science at Upsala Institute in Brooklyn and later professor of natural sciences and mathematics at Upsala College in Kenilworth, New Jersey, within commuting distance of Columbia University. The move was made because he wanted to take his Ph.D. with Britton. He registered as a candidate for that degree in 1895, held a scholarship at Columbia in 1895–1896, a fellowship in 1896 and 1897 and received his doctor's degree in 1898. On 14 January 1896 he was elected a member of the Torrey Botanical Club (*Bull. Torrey Bot. Club* **23:** 218. 1896). In that same year he published his *Flora of the Black Hills of South Dakota,* the result of a summer spent in that area in 1892 for the USDA (Rydberg, 1896b). The field work for the USDA was continued in 1896, when he collected again in Montana, this time accompanied by Julius Hjalmar Flodman of the Luther Academy at Wahoo.

The subject of Rydberg's dissertation was a monograph of the North American Potentilleae. Rydberg had already worked on *Potentilla* at Lincoln and was able to publish preliminary notes on the genus in June 1896 (Rydberg, 1896a). This work brought him into contact with Nelson of the University of Wyoming at Laramie (Williams, 1984). Nelson was already professor of botany and actively engaged in studying the flora of the Rocky Mountains. Rydberg had seen a publication by Nelson on *Potentilla* and asked him to send all available specimens of the genus on loan, mentioning that Wyoming seemed to be one of the richest states for the genus.

Nelson's biographer Roger L. Williams pointed out that "In view of Rydberg's immense importance in the history of western botany, botanical historians have lamented his failure to attract an adequate or sympathetic biographer and have noted that he suffered from the 'spiteful animosity' of some of his colleagues" (Williams, 1984). There was certainly some competition between Rydberg and some of his contemporaries, but this remained purely professional. Rydberg was too gentle a man to be spiteful but did not fear to express differences of opinion on scientific concepts.

The New York Botanical Garden started its field program in 1897. For its first expedition Rydberg was sent to Montana and Yellowstone Park. Funds were made available by William Earl Dodge, a member of the Board of Managers who was a philanthropist and patron of the sciences (Cox, 1903). The expenses of this trip, $672.14, were paid privately by Dodge and supported Rydberg in the field from June until September (Bull. New York Bot. Gard. **1:** 136. 1898). On this, his third trip to Montana, Rydberg was accompanied by Ernst Athearn Bessey, whose father had been Rydberg's Nebraska professor (Barnett, 1958). Apart from three weeks in Yellowstone Park, the two spent all their time in Montana. The Board of Managers, on receiving the report of the expedition on 2 November 1898, decided to publish Rydberg's *Catalogue of the Flora of Montana and the Yellowstone National Park,* a result of his three expeditions made in the summers of 1895–1897. The book (Rydberg, 1900) initiated the serial *Memoirs of the New York Botanical Garden,* still active today. Indeed, the present work is No. 58 in the series.

The correspondence between Nelson and Rydberg continued later in 1897. Nelson had sent undetermined specimens from Wyoming to the agrostologist Frank Lamson-Scribner at the National Herbarium in Washington, D.C., who forwarded the *Astragalus* to Rydberg. Rydberg thereupon suggested that Nelson himself could perhaps work out some apparently new species in that difficult genus "better than I" (Williams, 1984). This courtesy struck Nelson favorably, especially in contrast to the patronizing attitude of Edward Lee Greene. Notwithstanding this gentle behavior of Rydberg, Nelson remained wary. He was grateful "for the cordiality extended by the New York group" but "was already becoming anxious about an inordinate dependence upon them, especially if that dependence persisted as a tutelage," but he perceived a persistent aura of censorship. Nelson had resisted the exclusive claims to botanical wisdom assumed by Harvard and by Greene, and as Rydberg seemed more forcefully to advance his own views, Nelson sought alternative outlets for his new species.

Williams (1984) has well described the extreme sensitivity common at the time among American taxonomists. "Harvard" stood for the conservative tradition of Asa Gray, with broad species and genus concepts. Greene was the rebel of the West, rich in field experience but inclined to be as authoritarian as Harvard. The New York group must have appeared equally overbearing to the lonely Wyoming botanist. Nelson could not imagine that Rydberg might have shared his loneliness, notwithstanding the press of colleagues around him. So, while Nelson stayed on cordial terms with Rydberg and the Garden, he sent his manuscripts elsewhere. Nonetheless Rydberg continued to keep an eye on his friend in Wyoming. On one occasion he discovered that the name of a new species proposed in a manuscript submitted by Nelson to the *Bulletin of the Torrey Botanical Club* was a homonym and had to be replaced. The discovery was made when the article was already in proof and Nelson could not be reached at short notice. Rydberg renamed the species himself, but in a diplomatic footnote explained: "In his absence on a botanical expedition, I take pleasure in dedicating this species to the discoverer and describer": *Senecio nelsonii* Rydb. (Nelson, 1899).

In 1898 Rydberg was employed part-time by the Garden to work up the Montana collections. On 25 November he obtained his Ph.D. at Columbia, his dissertation being the aforementioned monograph of the Potentilleae, a substantial work of 223 pages and 112 plates (Rydberg, 1898). The drawings were made by F. Emil (a professional artist) and by Rydberg himself. His definitive appointment at the Garden, as assistant curator, came in June 1899. His first annual salary was $1080; it would climb to $5500 by 1930. The intermediate steps were $1920 in 1908, $2100 in 1909, $3000 in 1913, $3820 in 1920, $3900 in 1924, and $4500 in 1928. Rydberg was assistant curator from 1899–1906 and curator from 1907 to 1930 (Figs. 1, 2). All his publications in this period are listed by Barnhart in the *Journal of the New York Botanical Garden,* and a consolidated bibliography is published herewith.

The *Catalogue of the Flora of Montana and the Yellowstone National Park* (Rydberg, 1900) was published on 15 February 1900, a sizeable book of xi + 692 pages. The catalogue provides bibliographical and nomenclatural references for each species and cites the specimens consulted. Descriptions are given for the new species and taxonomic notes on critical ones. Several new species were contributed by his colleague Small. The total number of species and varieties recognized is 1976, of which 163 were new, and some 800 had been collected by Rydberg himself. These figures do not point in the direction of extreme splitting. Nelson, who reviewed the book, had great praise for it: "Nearly all the collections that have been made from the time of Lewis and Clark to the present seem to have been available for study . . ." (Nelson, 1900). However: "Al-

Fig. 1. Group photograph taken Aug 1916. From left to right: F. J. Seaver, M. A. Howe, G. V. Nash, J. H. Barnhart, P. A. Rydberg, C. A. Hollick, R. S. Williams. Courtesy of Archives, Gray Herbarium Library, Harvard University, Cambridge, Massachusetts.

together the work is so admirably done, and contains so much of value . . . one might be justified in wholly overlooking any inaccuracies. . . ." He added that a more conservative taxonomist would have recognized fewer taxa and regretted the application of the Rochester rules of nomenclature. This "resulted in the displacement of some familiar names, and even to one who thoroughly believes in the principle of priority it is a source of regret that it should necessitate so many changes." On the whole, however, the review was positive and Rydberg must have been pleased.

With the completion of the Museum Building at the Garden, Rydberg's private herbarium from Sweden as well as from the United States was added to the Garden herbarium (Small, 1900).

Rydberg was sent in July and August of 1900 to the Sierra Blanca of southeastern Colorado. He was accompanied on this trip by a volunteer assistant, Frederick King Vreeland; together "they obtained between 5000 and 6000 herbarium specimens, including a considerable number new to science" (Britton, 1900).

Back in New York Rydberg had occasion to write to Nelson concerning a paper submitted by the latter to the *Bulletin of the Torrey Botanical Club*. Lucien Marcus Underwood was the editor of the *Bulletin* but delegated the correspondence on phanerogams to Rydberg. Nelson's paper was the thirteenth of his series "New plants from Wyoming" (Williams, 1984). Among the new taxa was an *Asclepias* named in honor of Rydberg. The dedicatee, however, remonstrated that "it was in bad taste to name a plant even for a good friend when that individual had been neither its collector nor its describer." The species was renamed and published as *A. curvipes* (Nelson, 1901). The epithet *curvipes* was a somewhat surprising substitute for *rydbergii* in view of Rydberg's lameness. However, it was suggested by Rydberg himself (Rydberg to Nelson, 20 Feb 1901, University of Wyoming archives): "Please

Fig. 2. Photograph of P. A. Rydberg and Alfred Rehder, taken 1927. Courtesy of Archives, Gray Herbarium Library, Harvard University, Cambridge, Massachusetts.

let it have some other name. How is for instance *A. curvipes* from the strongly double-curved pedicel in fruit." Since Nelson and Rydberg had not yet met, Nelson could not have guessed the potential ambiguity of the epithet.

Rydberg also commented on Nelson's wish to describe a plant in honor of his wife: "I do not like to see anybody name a species after his wife. I am unmarried, but if I were married and thought ever so much of my wife, I would not do it. Heller was laughed at when publishing *Lupinus hellerae*" (Rydberg to Nelson, 26 Nov 1900, University of Wyoming archives). His advice was followed at the time, but 24 years later Nelson published *Calhounia nelsonae* (Nelson, 1924).

The summer of 1901 did not send Rydberg into the field but back to Europe for the first and last time. He was given leave of absence to visit a number of Scandinavian institutions and herbaria as well as Kew Gardens. His main charge was to arrange exchanges for New York, especially of herbarium material, but also of display material for the museum and of living plants. Nothing is known about his travels or visits to his family (Rydberg, 1901a).

Rydberg had enjoyed a thorough European classical training at his "Gymnasium." Schools of this standing provided advanced and sophisticated secondary training, based either on a religious or a humanistic theme. Rydberg was adept in Latin (as well as Greek) as shown by his scholarly paper in *Torreya* entitled "When in Rome do as the Romans do" (Rydberg, 1901b). He applied this proverbial exhortation to the formation of Latin honorific epithets. In biology two forms are used: a) the Latin genitive of the proper noun, and b) an adjective derived from the same. "Many botanists have agreed to use the former when the person in whose honor the plant is to be named has discovered it or done any other work in connection with it, and the latter when the author wishes to honor some friend without the latter being otherwise connected with the plant." The acceptance of this distinction was far from universal, however. Rydberg pointed out that the first procedure has its difficulties especially when following the rules of the Engler school at Berlin: "The practice of just adding -ii to a name ending in a consonant other than -i leads to classical absurdities such as '*Magnusii,*' from Magnus, a name already in Latin form." Rydberg went into great detail, proposed some changes, and stated that "The latinizing of proper names is a matter that needs the attention of a botanical congress." His suggestion was actually taken up, not only at the Vienna Congress of 1905, but at all other botanical congresses of the twentieth century.

The fall of 1901 was good to Rydberg. In the first place he obtained a raise in his salary to $1200 per year, and he could furthermore look back at no less than eleven publications, including an important contribution to Britton's *Manual of the Flora of the Northern States and Canada.* This *Manual* was an updated digest of Britton and Brown's *Illustrated Flora.* Rydberg inserted quite a few of his new species in Britton's *Manual,* by means of which it was possible "to name any plant from adder's tongue to blessed thistle" (Coville, 1901). All botanists at the Garden were likewise forging ahead in what was called by a contemporary magazine "this new Jerusalem of botanists" (*Success,* Oct 1901). Rydberg's main initiative in 1901 was a *Flora of Colorado,* published in 1906.

At this time Coulter and Nelson were planning a flora of the central Rocky Mountains to replace Coulter's *Manual of the Botany of the Rocky Mountain Region* (Coulter, 1885), a book frequently used by Rydberg in his early days. The *New Manual of Botany of the Central Rocky Mountains* was to come out in December 1909, but the authors knew as early as 16 September 1901 (Williams, 1984) of Rydberg's plans for a flora covering Colorado alone (rather than Colorado and Wyoming together). There is no trace of any official plan in that direction among the documents of the Garden. However, better quality books often come as a surprise and not heralded long in advance. Rydberg in fact emulated his director: both just pushed ahead, worked hard on their major works, and made practically no noise about them.

The years 1902–1905 were given mostly to the writing of the *Flora of Colorado.* Rydberg extended his correspondence with botanists living in Colorado such as Nelson and Theodore Dru Alison Cockerell, an all-round naturalist who taught biology at Las Vegas, New Mexico in 1902–1903, was curator of a museum at Colorado Springs in 1903–1904, and from 1904 was connected with the University of Colorado at Boulder (Weber, 1965). Correspondence with Nelson was by far the most frequent. Relations between the two remained cordial, even though Nelson continued to drift away from his orientation to New York. He was dissatisfied with the way his papers were treated by Underwood, editor of the *Bulletin of the Torrey Botanical Club.* In the summer of 1903 he wrote to Rydberg of his concern (Williams, 1984). Rydberg replied on 4 August that "The censorship business was an idea of Professor Underwood, which I never liked." The situation changed, however, in 1903, when Barnhart succeeded Underwood as editor. Rydberg was eager to keep in touch with Nelson also because of the isotypes of new species sent by him to the Garden. He again offered his assistance, "but if you regard it as a censorship, I do not like to continue the former relations. It is true that I have hindered you from publishing some good species [sic], but there are also cases where I have saved you from republishing old ones as new." The relations remained friendly and the exchange of specimens continued. By 1910 the Garden had sent 2806 specimens to Laramie and received 2245 in return (Williams, 1984).

Nelson tried to keep the peace on all fronts. The Harvard botanist Merritt Lyndon Fernald, assistant to Benjamin Lincoln Robinson at the Gray Herbarium, wrote to Nelson warning against premature publication and urging comparison of presumed new taxa with the older collections such as were kept at Harvard. Fernald objected strongly to the "American Nomenclature" of the Britton school. The botanists at Harvard were strongly Kew-oriented and hence preferred established custom to clear-cut rules based on general principles. Nelson realized that, after all, he was "closer to Rydberg than to the gentlemen at the Gray" (Williams, 1984).

The relationship between Rydberg and Nelson was consolidated when they met in person. During the summer of 1905 Rydberg went again into the field, this time mainly in Utah. On his way west he visited Nelson at Laramie. Williams reports that "while the encounter was amicable, Rydberg managed to convey criticism of Nelson's association with Coulter. He spoke freely about his forthcoming *Flora of Colorado* and left no doubt that it would soon be in print" (Williams, 1984). Coulter, who was immediately informed by Nelson, reacted to Rydberg's criticism that Nelson had published too many new species "that they are but a drop in the bucket compared with the number he is publishing. Of course you are all making blunders. I know this, because I was always making them years ago. . . ." Apparently Coulter, who had turned from taxonomy and floristics to teaching and administration, regarded "splitting" as a juvenile phenomenon. Rydberg referred to his meeting with Nelson in his official account of the trip: "On my way I stopped for a day at Laramie, Wyoming, to make a visit to the State University. I spent a very pleasant time with Professor Aven Nelson and saw nearly all of the types of his new species, of which there are not cotypes [isotypes] in the herbarium of the Garden. This will help me considerably in my work on a manual of Rocky Mountain Botany" (Rydberg, 1905b).

In 1902 Rydberg was elected a member of the Botanical Society of America, then still a small club with restricted membership.

On 11 November 1903 Rydberg married Alfrida Amanda Rydberg, the daughter of one of his cousins. From this marriage sprang four children, of which the oldest daughter died when 14 years old. Mrs. Rydberg and her three other children survived their husband and father. The family was closely united and linked together especially by the Lutheran church, in which Mrs. Rydberg was very active.

An interesting sidelight on Rydberg in those years is his review, in *Torreya,* of a monograph of *Campanula rotundifolia* and its allies written by an Austrian author (Rydberg, 1903; Witasek, 1902). Rydberg was full of praise for this revision of a still controversial species-group. The taxonomic details are hardly relevant here but the general comments are noteworthy. "Even the American species are treated in a way very unlike the unsatisfactory one in which Europeans usually monograph American plants. If this monograph is compared with that of *Oenothera* by Léveillé [1902] its superior quality is evident."

> *After I had glanced over the 106 pages of the text and studied what was of most interest to me, especially all that related to American botany, I turned to the preface and here awaited me the greatest surprise. The author is a woman. On the title page the author's name is given as J. Witasek without any title whatever, and in the text the personal element is as it ought to be so eliminated that there is no indication of the gentler sex. Only the first line of the preface contains the words 'Verfasserin' followed by a few 'die' and 'ihre'. Not that I believe a woman incapable of a good piece of work, far from it; but in Europe there are but few women that receive a university education and besides their education is generally very unlike that of men. Therefore, the monograph indeed is a credit to both the author and her sex. . . .*

Rydberg was not so well acquainted with Richard von Wettstein, the godfather of Witasek. If he had known the situation at Vienna at the time he would have seen that female collaborators were not so exceptional. Witasek was a teacher at a girl's high school in Vienna and worked, as a volunteer, with Wettstein from 1899 to 1911. Women biologists were not at that time so scarce in Europe as Rydberg supposed. A well known contemporary was Johanna Westerdyk, who worked with Karl Goebel in Munich from 1904 to 1906 and took her Ph.D. degree with H. Schinz in Zurich. Westerdyk was Dutch and served as a professor of phytopathology at Utrecht from 1917 on.

After a few years of hard work but relatively few publications, at least by his own or Britton's standards, 1905 saw Rydberg begin a new stream of printed matter. Rydberg contributed to the first fascicle of the *North American Flora,* which came out on 22 May of that year, five pages of Penthoraceae and Parnassiaceae. In the second fascicle published on 18 December, he collaborated with Small on the account of the families Saxifragaceae and Hydrangeaceae (Rydberg, 1905c, 1905d; Small & Rydberg, 1905a, 1905b).

The summer of 1905 again saw Rydberg in the field, from late May until August. Collecting was centered on Utah, where he was accompanied by E. C. Carlton, a professor at Gustavus Adophus College, St. Peter, Minnesota, another American-Swedish Lutheran. Because of the season, the exploration of Utah was concentrated on the mountains. The two men travelled as far as possible by railroad and changed then to wagon and horses. Notwithstanding his lameness Rydberg climbed several mountains ranging from 10,000 to 12,000 feet (Rydberg, 1905b). He made a side trip to San Francisco to visit the California Academy of Sciences where he met the curator, Alice Eastwood, and former curator Mary Katharine Brandegee. The purpose of this visit was to study material of Saxifragaceae for the treatment of the family in *North American Flora.* He returned to New York with a harvest of 5126 specimens from Utah and a few from California.

From Henry Allan Gleason's autobiography and thumbnail sketches of botanists (unpublished manuscripts, NYBG archives) we have an impression of Rydberg in 1905:

> *Right around the corner [of the herbarium room] sat a big blond, lame Swede, with a high-pitched voice, who stuttered when he spoke English. That was Rydberg . . . Per Axel Rydberg was a big man. He had the complexion of the typical Swede, and when I first knew him he had a golden Vandyke beard and yellow hair, but these turned to gray in his later years. He had been educated in Sweden for a mining engineer, but was crippled by an accident and had to give up that profession. He then turned to botany, in which he had apparently long been interested. . . . He was a great taxonomist. Measured by familiarity with western plants, by acquaintance with the literature, by painstaking study, he ranks well up in the forefront of the taxonomists of the first three decades of this century. He limped when he walked, and one foot audibly dragged along the floor at every step, but that did not deter him from much hard field work in the Rocky Mountains. He stuttered when he talked in his high and rather squeaky voice; he never lost his foreign accent, and it was rather painful to listen to him talk. He did not stutter when he spoke Swedish.*
>
> *Like all other taxonomists of the time, he was a botanist of the old school. He did not believe in varieties. If a plant was worth describing, it should be a species. There was no use in cluttering up nomenclature with trinomials. He believed in consistency. If two genera were separated by characters of some particular nature in one family, then the same characters should be valid to separate [illegible]enera in every other family. When he was revising [illegible]*ragalus, *he poured out to me all his taxonomic [illegible]es. That was because my table was close to [illegible]would show me fully accepted genera in the Legumes which were distinguished by certain features of the pod, and would show me species of* Astragalus *in which the same difference existed. All his segregates of* Astragalus *were based primarily on this sort of reasoning.*

A first general review of *Astragalus* and its segregates came out in January 1906, a forerunner of his ultimate treatment of the genus for *North American Flora* (Rydberg, 1905a, 1929a, 1929b, 1929c).

The *Flora of Colorado* came out in the summer of 1906, published by the Agricultural Experiment Station at Fort Collins, Colorado. The work goes back to the systematic collection of Colorado plants made by James Cassidi, professor of biology at Fort Collins. This collection, like those of Charles Spencer Crandall and Jacob Hover Cowen, is now in the herbarium (CS) of the Department of Botany, Colorado State University at Fort Collins. The College and New York Botanical Garden had agreed to collaborate on a catalogue of the collections. The outcome of the agreement was the *Flora of Colorado* which is, in fact, like the *Flora of Montana,* an annotated catalogue providing names and synonyms, analytical keys to genera and species and for each of the latter a list of localities, but no descriptions. Rydberg used the annotations of Crandall, also a professor at Fort Collins, who had in manuscript a list of local collections. Crandall adhered to the "Gray school" which meant very conservative taxonomic and nomenclatural concepts. In a review of the *Flora of Colorado* Nelson (1906) noted "That the preparation of the manuscript could not have fallen into abler hands needs not to be stated but it is the irony of fate that the work begun by a conservative should have been revised and concluded by an ultra-radical of the recent school." In the preface Rydberg mentioned that the original manuscript contained about 1400 species, his own final text included 2912 species in 700 genera. This was not due simply to a narrow definition of species but mainly to the circumstance that Rydberg had access to the collections at the Garden and elsewhere that Crandall had not seen. The general sequence adopted in the *Flora* was that of Engler and Prantl, but, as Nelson noted, ". . . In the genera recognized there is considerable deviation from that work." *Astragalus* had been broken up into 17 genera, but "a very difficult piece of work had been done remarkably well" (Nelson, 1906).

Rydberg was grateful for the review and wrote so to Nelson, but added "Don't you think, however, that 'ultra-radical' is rather strong, what shall we then call Dr. Greene, or even yourself?" (Williams, 1984). Nelson received supportive comments from other colleagues. Among them is that of the Boulder naturalist Cockerell, who seemed to have been annoyed by the way Rydberg had treated several taxa, and noted: "I find Rydberg a very difficult person to cooperate with in any way. He seems to get angry and shuts up like a clam at the slightest suggestion he is wrong! The New York people are doing splendid work, but they are much too inclined to hold the view that 'orthodoxy is my doxy, and heterodoxy is the other fellow's doxy' and this, after protesting so much against the oldtime 'tyranny' of Gray!" (Williams, 1984). Cockerell and Rydberg were in some respects two of a kind. Questions were also raised as to the *Flora*'s usefulness. Nelson could work with it, but one of his later correspondents and students, Edwin Blake Payson, who had used the book in the field, wrote to Nelson: "It often seems impossible to understand Rydberg" and "I do not try any more." However, Payson's taxonomic experience was rather limited at the time. It was agreed that the *Flora* was not an "easy book" (Williams, 1984). In 1910 Greene, an experienced critic, had nothing but good words for Rydberg's thorough treatment, praising especially its excellent general coverage and its narrow species concept. He pointed out that Nelson's competing *New Manual of Botany of the Central Rocky Mountains* of 1909, written with Coulter, though it covered an area three times that of Colorado, listed only 2733 species, compared with Rydberg's 2912 species for Colorado alone (Coulter & Nelson, 1909; Greene, 1910). Nelson had reduced 400 of Rydberg's species, but had also placed in synonymy names proposed by Asa Gray, the man "... whom I hear people speak of as having been Dr. Nelson's botanical father." Greene was especially annoyed by the fact that "... the multitudinous Nelsonian species are commonly maintained as valid. I doubt very much that any botanical community on either side of the Atlantic will be found to be of the opinion that Dr. Rydberg and I are the reckless species makers, and that Dr. Nelson is the careful, cautious and discrete conservative. I ... have no such opinion of my friend in Wyoming, nor, as I said, do I believe it exists anywhere." Greene's outspoken opinions must be taken with a grain of salt, but it is clear that in this case he recognized the injustice of the proposition that "my splitting is not your splitting."

The publication of the Coulter-Nelson *New Manual* gave rise to further criticism of Rydberg, especially among the proponents of a wider species concept. Theodore Christian Frye wrote to Nelson from the University of Washington: "I am glad you cut in on Rydberg in your *Rocky Mountain Flora*. ... The tendency to split species ought to be stopped. It is doing more to bring botany into disrepute than any other thing I can think of " (Williams, 1984). This was of course a two-edged sword, equally applicable to Nelson.

A severe critic of Rydberg's *Flora* was Marcus Eugene Jones, a capable botanist accustomed to voicing extreme, sometimes scurrilous opinions of others (Lenz, 1986). Jones wrote about Rydberg's *Flora* in his *Contributions*: "This handy Volume by Rydberg is on the whole the best thing that has ever been printed on the Colorado flora and cannot fail to be helpful" (Jones, 1910). However, he criticized Rydberg for confining himself to the material of Columbia University and ignoring the specimens at the Gray Herbarium as well as those at the United States National Herbarium in Washington, D.C. Rydberg was, of course, a "splitter," and his book "... fairly bristles with fake species and genera, and yet his keys are remarkably well made and valuable though they are at times misleading. If he could divorce himself from the species making craze his work would be good" (Jones, 1910). It is true that at that time Rydberg had not made extensive use of the herbaria at Harvard and Washington; however, he had seen much of the old material at Kew, and the Columbia herbarium contained many duplicates from Kew, Harvard and the National Herbarium. For his 1917 *Flora of the Rocky Mountains* Rydberg took Jones's remarks to heart and frequently visited the two latter herbaria. In a much later publication Jones gave his general opinion of Rydberg in even stronger terms:

> *Rydberg was a kindly Christian man, well liked by those who knew him best. He was capable, sloppy and a mule. ... He was almost as bad a splitter as Greene, characterizing most of his species on single specimens without taking into consideration well known ecological factors, and without attempting to collect the intergrading material in the type localities of his proposed species. He had no*

conception of genus, and a very poor one of species. . . . (Jones, 1934)

It was a harsh judgment, neither quite fair, nor correct. The "splitting" was of a different nature, as we shall see when discussing his work on *Astragalus.* Rydberg's splitting was in general unpopular in the West. However, towards the end of his life the splitting became much less extreme, as is shown by his last and probably best work, the *Flora of the Prairies and Plains of Central North America,* published shortly after his death.

It is also of importance to point out that Rydberg's "splitting" and "taxonomic radicalism" was in no way forced upon him by Britton, as has frequently been surmised. Britton was interested primarily in "production": his staff should publish, preferably at a great pace. It was not "publish and be damned." He himself had a consistent program of regular publication, avoiding time-consuming details. There is no direct indication that he exercised any control over his staff's taxonomic opinions. Rydberg's taxonomy was definitely independent, though it may have been influenced, but not imposed upon him, by the taxonomic philosophies of his colleagues at the Garden. This could explain his early recognition of varietal names, e.g., in his Sand Hills flora (1895c), but only binomials in his later works. He had a somewhat rigid belief in "differences," a sharply analytical mind, and an ignorance of variability in populations, whether genetically induced or caused by environmental factors. There is also no indication that Britton had the time or the inclination to read or critically study Rydberg's manuscripts before they went to press. He had learned from Rydberg's dissertation that he was an acute observer and from then on left him completely free to formulate his own opinions and concepts. Britton was actually against "this segregation business," which "must be stopped," but he also believed in academic freedom. It has sometimes been suggested that Britton did not financially support the publication of Rydberg's books because he did not share Rydberg's opinions. This was definitely not the case. Under Britton there was a general policy at New York, applied also to Britton himself, that the author had to take financial responsibility for books. This had nothing to do with censorship. This policy was discontinued while Elmer Drew Merrill was director of the Garden.

With the reorganization of the staff of the Garden in 1906, Rydberg was promoted to the rank of Curator, together with Howe, and Small became head-curator. The curator's salary became $1920 per year. The most important administrative change, however, was the promotion of William A. Murrill, so far a junior member of the staff, to the rank of first assistant, comparable to that of vice-president in the nineteen-eighties. The position carried a much higher salary but involved participation with Britton in administrative duties and was certainly not an undiluted blessing.

On 23 May 1907 exercises commemorative of the two-hundreth anniversary of the birth of Linnaeus were held in several New York institutions. The Swede Rydberg was a natural choice to give one of the addresses. He spoke in the afternoon at the Garden on "Linnaeus and American botany." He made the points that Linnaeus became acquainted with several American plants, e.g., *Rhus toxicodendron,* poison ivy, as early as 1732, that he had studied American plants in Holland between 1735–1737, that he had sent his student Pehr Kalm—a collector commemorated by the genus *Kalmia*—to the New World, and stated that Linnaeus knew and described nearly one hundred species from the United States and Canada in the second edition of the *Species plantarum* of 1762. Rydberg would not have been Rydberg if he had not adorned the occasion by publishing a new species: a fourth species of the genus *Linnaea,* from Alaska. On the same day the Linnaeus bridge over the Bronx River between the Garden and the Zoo was dedicated as a permanent memorial to the link between botany and zoology and a commemorative plaque was unveiled on this recently completed stretch of what is now Fordham Road.

In the meantime Rydberg continued his work on the Rosaceae for the *North American Flora.* Two parts, comprising no less than 246 pages, came out on 12 June and 20 November 1918. In the long run Rydberg became a major contributor to the *Flora.* In all he treated twenty families, among them large ones such as Rosaceae and Fabaceae, and several tribes of the Compositae. While this had become expectable and was hardly noticed in the contemporary botanical press between 1905 and 1927, his regional floras aroused extensive comment and controversy.

In the period between the *Flora of Colorado* of 1906 and the *Flora of the Rocky Mountains* of 1917, Rydberg was preoccupied with the new flora and *North American Flora.* The work on the Compositae for the *North American Flora* was very time-consuming, even though the first fascicle of vol. 34, Helenieae, ultimately took up only 80 pages. Notable in this fascicle was the contribution by Harvey Monroe Hall of two genera. Notwithstanding the differences of opinion between them on the delimitation of taxa, Hall and Rydberg's relationship was good at this time. In this period Rydberg frequently visited other major centers of taxonomic research: Harvard, Washington, D.C. (National Herbarium), and Philadelphia (Philadelphia Academy of Sciences); these visits sometimes lasted several months at a stretch. In 1911 he had an opportunity to go into the field again, to Utah, this time with Albert Osborn Garrett, a high school teacher from Salt Lake City (Rydberg, 1911). He was away from New York from 10 June until 18 September, being in the field for a week in Minnesota, and, with Garrett, in southeastern Utah for most of the rest of the time. He also visited local herbaria in Colorado, Nebraska, and Missouri. He brought back some 6000 collections representing 1740 numbers of phanerogams and a small number of cryptogams. According to his own estimate he had seen and collected material of some 1000 species.

The *Flora of the Rocky Mountains* appeared in February of 1918, the fruit, as noted in the preface, of "more than twenty-five years of study of the flora of the Rocky Mountain region," and seventeen years after the description was drawn for the book. It was a monumental volume: 1121 pages of text in all, in which 5897 species were treated in 1038 genera (Fernald, 1936; Rydberg, 1917).

The most important review of the book came from Paul Carpenter Standley (1918), then botanist at the National Herbarium in Washington, D.C. Standley, at 34 years of age, was a mature and accomplished taxonomist, very knowledgeable on the subject.

> *The Rocky Mountain area,* he wrote, *is a striking example of insularity in botanical work. Too often isolated individuals, actuated by the best of motives, but frequently without the assistance of adequate literature or herbaria, have carried forward their work independently of that done in other centers. Unwarranted importance has been assigned to geographic isolation as a factor in determining specific distribution, with the result that a given species has sometimes been described several times from separate localities. If this* Flora *were no more than a correlation of nomenclature it would serve an invaluable purpose; but it is far more than a 'nomenclator'. . . . No botanist is so well qualified as Dr. Rydberg to prepare a manual of the Rocky Mountains.*

Standley mentioned the author's wide field acquaintance with the plants through several seasons as well as his numerous other publications on the subject. He mentioned incidentally that about one thousand species carry Rydberg's name.

> *Dr. Rydberg's ideas concerning species and genera are well known to the botanical public, and while there are many who will not agree with him in regard to the limits of groups, especially genera, such differences of opinion will not detract from the general usefulness of the flora. Synonyms are cited freely to coordinate generic segregation as well as variations in usage under different codes of nomenclature. If a botanical work is fairly and accurately written, the code of nomenclature followed and the measure of specific and generic limits adopted are matters of minor importance; and there is no doubt that the present work has been so prepared. Of special importance, too, is the fact that the descriptions have not been compiled but are drawn direct from the plants themselves.* (Standley, 1918)

Standley praised the brief but adequate descriptions which supplement the key characters and the wealth of information concisely presented. In summary he stated that the *Flora* is a "volume that will be indispensable for many years, and will long stand as a monument to the industry, enthusiasm, and discriminating vision of the author."

Nelson did not review the book. He tried to persuade one of his former students, James Francis Macbride, to write a review for the *Botanical Gazette.* Macbride, then only 25 years old, refused and insisted on a review by Nelson. The latter replied "You have convinced me that the botanical world is waiting breathlessly for my opinion of Rydberg's *Flora.* I will try to relieve the strain quite soon. . . ." Nothing, however, came of it. Nelson may have been reluctant because he had reduced 400 of Rydberg's species to synonyms in his *New Manual,* or he was perhaps over-committed, having been recently appointed president of the University of Wyoming (Williams, 1984).

The sharpest criticism, not unexpectedly, again came from Jones, not the most gentle of gentlemen. His review is an example of the heat that may be generated by a simple flora:

> *This book is what I have in the past dubbed bughole botany, species depending on the number of bug holes in the leaves. Rydberg has no conception that there is any such science as ecology, or that environment has any effect on species variation. . . . In the matter of names, Rydberg is very sloppy. . . . When we look at this last botanical abortion of Rydberg . . . we shudder at what the next edition will be like. . . . In the matter of specific distinctions I can see little difference between Rydberg and Fernald. Both are splitters from Splitville, though Rydberg glories in wallowing in the mire. While Fernald does not seem to know that he is in it.* (Jones, 1929)

With passage of time, Standley's review appears the more balanced and objective. He had no axe to grind, as was the case with Jones.

A second edition of the *Flora* came out on 21 January 1923. The bulk of it consisted of the original sheets of the 1917 edition (Fernald, 1936). The stock of bound copies had run out in 1922 and the remaining sheets were bound with 34 new pages of additions and corrections. This 1923 issue was reprinted as a whole in 1954 and again in 1969. The book was evidently in good demand for half a century and, notwithstanding adverse criticism and some ageing, remained a useful, often consulted source of information.

Rydberg turned sixty years old on 6 July 1920, and spent most of the last decade of his life working on the *Flora of the Prairies and Plains of Central North America* as well as, understandably, on the *North American Flora,* especially the Leguminosae, and in particular on *Astragalus.* He made only three further field trips. In 1925 he travelled in Virginia and North Carolina, this time by automobile, accompanied by a student from Massachusetts Agricultural College, John Tuttle Perry (Rydberg, 1926a, 1926b). The trip lasted 10 weeks, partly as summer vacation, Britton having given him an extra month to collect for the Garden. It yielded, as always, a large set, 3000, of herbarium specimens. A fully professional trip was made in 1926 to the area of the coming flora of the prairies and plains, which included visits to herbaria as well as field work.

Rydberg also took time off to attend the International Congress of Plant Sciences, actually the fourth International Botanical Congress, held at Ithaca, New York in August, 1926. It was at this Congress, Rydberg's first and last, that he met several of his adversaries, of whom the most notable was Harvey Monroe Hall from California. The Garden was well represented at the Congress by Rydberg, Howe, Arlow Burdette Stout, Barnhart, Fred Jay Seaver, Charles Arthur Hollick, and Gleason, supported by $300 towards expenses provided by Britton. This sum was given to the group, not to each person. Gleason mentioned the meeting in which Rydberg and Hall confronted each other (Gleason, autobiography, NYBG, archives):

> *There was just one episode* [at the Congress] *that left me with a tinge of regret. Hall and Clements had just finished their studies of* Artemisia *and Hall read a paper on their viewpoints, methods, and general concept of the species. Rydberg had also recently published on the same plants and his ideas and concepts were radically different. Hall criticized Rydberg severely. Hall, now at the height of great personal popularity, genial and urbane, tall and handsome, fluent and clear in speech, easily carried the audience with him. Then Rydberg had the opportunity to reply. He dragged his lame leg as usual* [and spoke with] *falsetto voice, in his usual stuttering words. He could not have swayed that audience to agree with anything that he might have said. If Hall had said that two and two were five, the audience would have believed it; if Rydberg had proved that two and two were four, the audience would have scoffed. I felt sorry for Rydberg then and I still feel sorry for him today.*

Hall influenced plant taxonomy in many ways, not least in the development of what later was called "biosystematics," the application of genetics and ecology to fundamental plant systematics. In 1923, with the ecologist Clements, Hall had published *The phylogenetic method in taxonomy,* an influential and trend-setting publication for modern plant systematics. His criticism of Rydberg, however, was only partly justified. Rydberg defended his stand at the Ithaca Congress in an address entitled "Scylla or Charybdis," which contains a clear-cut account of his taxonomic philosophy and a defense of his narrow species concept (Rydberg, 1929d).

"We have seen the endless and infinitesimal splitting into categories of the German School." This had led to such names as *Saxifraga aizoon* var. *aizoon* subvar. *brevifolia* forma *multicaulis* subforma *surculosa,* much more a statement of classification than a name. "... For twenty years [we] have been thoroughly disgusted with their absurd nomenclature of trinomials, quadrinomials, and polynomials. We [that is also Rydberg] believed that any form that is worth describing is worth a name and that a binomial is better than any other. But we might have forgotten that there is a Scylla as well as a Charybdis." Hall and Clements criticized "the tendency... of segregating genera and raising their sections to generic rank" and Rydberg replied "It is true that the splitting may have gone too far, that the school of taxonomists represented by my own institution has dared too near to Scylla. Some may think that we are already shipwrecked." The technical discussion, even by itself amazing and enlightening, cannot be repeated here, but it is still worth reading, sixty years later. In principle, the dilemma—the choice between "lumping" and "splitting"—is still with us today.

Rydberg agreed fully with the biosystematists that "evolutionary taxonomy must regard the segregates of the last 25 years [i.e., many of his own narrow species] chiefly in the light of tentative proposals as to the course of evolution." As long as the biosystematic research is incomplete the species described by him and others "have to remain tentative. I leave it to others to find out whether the characters, prominent or trifling as they may be, on which they are based, are constant enough to place them as distinct species." It is clear that in half a century since Rydberg's remarks this has not yet been achieved. In fact biosystematic research is so time-consuming that it produces examples and suggestions but no definitive, universally applicable solutions.

> *I think that I have done a service to taxonomy if I have pointed out existing characters. I also think that a form that is worth describing also should have a name. I prefer a binomial. When I use a binomial I do not pretend it means the same as a species in the Hall-Clements style. I do not pretend to be able to pass on its rank, because in most cases it is impossible.... It is not a craze for species-making, nor yet for attaching my name to a plant name that induces me to propose species.*

Rydberg found discrepancies between the number of actual experiments, the elaboration of methods, the work done, and the results. Here he pointed out a fundamental fact, discriminating between the work of the pioneer who describes diversity, not of a small specially chosen group of plants but of an entire flora, and the biosystematist's more elaborate, narrowly focused methodology. Synthesis of the two approaches is slow and criticism of the pioneer effort in the interval between gathering the facts of diversity and its explanation, is ill-considered and unjust. Rydberg quoted a prominent ecologist's statement: "One of the noblest aims of ecology is the destruction of many of the 'species' of our manuals" (Cowles, 1908). If the ecologist had stated "that the aims should be to find out the truth, even if it should cause the destruction of many of the species, I would subscribe to it." Rydberg questioned whether he himself had been seized by the monster Scylla, but questioned also whether Hall and Clements had successfully steered the middle passage. Every botanist presuming to judge the merits of Rydberg's taxonomic work should read his "Scylla and Charybdis." It is well written, and after sixty-odd years it is still as relevant as in 1926. He concluded:

> *For years I have admitted that I am a splitter, but I have prided myself on being consistent, as, for instance, in splitting some genera so as to make them equivalent to other genera in the same family or tribe, but I have just discovered that in order to save already described species of other botanists, I have added some of my own of the same category and made my species concept very inconsistent. I am inclined now to the slogan "Darn consistency."*

In writing Rydberg was persuasive and disarming; his difficulties at the Congress were, however, a stuttering, restricted command of spoken English and a constitutional modesty. He did not cut much of a figure at Ithaca. But congresses and lectures are ephemeral, while the written word endures.

The above discussion is applicable further to Rydberg's work on the genus *Astragalus* as widely defined. He published many preliminary papers bearing on classification of *Astragalus* but his summary exposition is the treatment of the "Astragalanae" in *North American Flora* (Rydberg, 1929a, 1929b, 1929c). Barneby (1964) discussed Rydberg's work in the introduction to his *Atlas* and offered a contemporary definitive evaluation of the nature and importance of Rydberg's work in this field. "It is the fashion nowadays, I think, to underestimate Rydberg's ability and to belittle his achievement. His narrow and severely technical concept of genus and species is at the opposite pole from the modern view based on evolutionary and genetical principles. But Rydberg, too, had his principles, and they were firmly based on logic." Barneby referred to "Scylla and Charybdis," an

> *urbane* apologia pro vita sua, *which is a confession of faith, in no sense dogmatic, and completely disarming. . . . There is an exploratory phase in the description of all floras, when taxonomy, strictly speaking, is hardly possible, and Rydberg presided over that period in the botanical history of intermountain and Rocky Mountain United States, the parts of North America where* Astragalus *is most highly developed. Rydberg's* Floras *and revisions of genera are in reality little more than illustrated catalogs and for this purpose the binomial was the handiest possible label. It should never be supposed that Rydberg was unaware of the protean aspects of the species. . . . Rydberg had few peers; perhaps too often he was inclined to leave to a succeeding generation the possibility for assessing the meaning and value of the characters observed. . . . Rydberg was in the first place a scholar in the European tradition, and he had comparatively limited interest in or experience with living plants.*

Barneby is also critical of Rydberg: the "austere simplicity" of his keys especially "is no match for the complex realities to which it appears but fails to introduce the reader." Rydberg described 130 new Astragali in various genera; Barneby reduced 80 to synonymy "and of the surviving 38% much more than half are degraded to varietal rank." In this respect the field botanist Marcus E. Jones comes off better: of the 150 species proposed as new by Jones 63 were treated as synonyms, "but much more than half of the remaining 60% are retained in their original category."

In addition to these notes in his *Atlas,* Barneby, in a letter to Frans Stafleu, dated 26 April 1986, has given a general opinion of Rydberg as a taxonomist, with some additional notes on the generic splitting.

> *Rydberg was a supremely able observer. One can compare his descriptions with the material from which he derived these and find every detail exactly observed and recorded. His cast of mind was logical, and I think his Achilles' heel was the infeasibility of his logic. His generic splitting in* Astragalus *started in a small way, with stress on relatively gross differences in pod-structure. As new discoveries blurred what first appeared to be significant discontinuities, the only way to save the system was to propose finer and even finer segregates. But the system is perfectly logical. The trouble is that he was either unaware of the prevalence of parallelism, or thought it irrelevant to a system of classification. I would say that one can rely with complete confidence on Rydberg's honesty of purpose—his scientific integrity, and that the charge (current among envious contemporaries) that he wanted above all to see his name associated with as many plant species as possible, reflects greater discredit on the accusers than on the accused. All letters that I have seen in Rydberg's hand are models of gentle courtesy. His life was, I believe, simple, virtuous, and possibly austere (like that of Bentham not good biographical material). His life-story must be read in his floras and revisionary papers, and there is a tremendous body of solid accomplishment there. People tend to recall excesses and mistakes and pass over as though of small account what was really, at the time, original and lasting. Have you happened on Rydberg's "Scylla or Charybdis?" It speaks better than I can of Rydberg's botanical philosophy. As a field botanist Rydberg was nothing much. His physical impairment must have contributed to this, but the lack of field-experience doubtless colored his taxonomic assessments.*

On 13 June 1929 Rydberg left New York for a field trip to Kansas and Minnesota which was to last three months. We have no report on this trip, which came to a premature end. On 22 August 1929 he wrote to Howe, the interim director of the Garden (NYBG archives): "I hurt my right foot Wednesday. Originally it was a slight muscle sprain and chaffing of the shoe. I paid no attention to it and made a long tramp the following day on an island off Great Portage, Minnesota. The foot was sore in the evening and somewhat swollen. Friday morning it was worse and I walked but little that day." Since there was no improvement he returned to Great Portage, then a small village without a doctor. He ended up obtaining medical treatment in Minneapolis, had some rest, and decided not to go to Kansas but to return to New York. After all, he was 69 years old. His companion on this excursion was Carl Otto Rosendahl of the University of Minnesota, himself a cripple who could not stand upright.

On 10 December 1929 Howe wrote to Britton (NYBG archives), who had started his retirement with a trip to Puerto Rico, that Rydberg had had a sort of collapse, possibly aggravated by a cold, but reported on 14 December (NYBG archives) that he was back at work. Britton had resigned his directorship on 1 August 1929 but Howe continued his habit to report the goings on at the Garden during Britton's absence. The new director, Elmer Drew Merrill, was expected to assume his duties on 2 January 1930.

The *Flora of the Prairies and Plains of Central North America* was now reaching completion. During the Britton period "books" of staff members either came out as *Memoirs,* paid for by the Garden, or were published independently at the financial risk of the author. This policy was abandoned by Merrill. The Scientific Directors decided on 14 June to publish Rydberg's new flora as an independent publication of the Garden. On 13 December 1930 they determined in principle that all publications by staff members would henceforth be issued by the Garden. The estimated cost of producing Rydberg's book was $7000, and the profits from sales were to be shared between the Garden and the author.

On 14 January 1931 Howe wrote to Britton that Rydberg was laid up again (NYBG archives). From then on Rydberg came to the Garden only at irregular intervals, the last time only eight days before his death, which occurred early in the morning of 25 July 1931, at his home, 3304 Rochambeau Avenue, in the Bronx. Howe, in a brief memorandum (NYBG archives) mentioned that "even in the afternoon before the fatal morning, the writer of the present note received from him at his bedside clear-minded instructions as to the completion and printing of his 900-page *Flora of the Prairies and Plains of Central North America* of which he had read about one half of the proof sheets."

Howe saw the book through the press. It was published on 14 April 1932, a book of 969 pages, treating 3988 species (Rydberg, 1932). The preface was written by Howe, who explained that Rydberg had had in mind a complete manual of the Spermatophyta and Pteridophyta of the states of Kansas, Nebraska, Iowa, Minnesota, South Dakota, North Dakota, and of southern Manitoba and southeastern Saskatchewan. He pointed out that in this flora Rydberg followed, for the first time, the International Rules of Botanical Nomenclature adopted at Cambridge, England in 1930 rather than the American Code. After the nomenclatural peace treaty of Cambridge the American Code had served its purpose and lost its importance. The book was well received, even though Rydberg had adhered, on the whole, to his narrow species concept. Alfred Barton Rendle, formerly an adversary in the international debate on nomenclature, wrote a sympathetic review (Rendle, 1933). "A melancholy interest attaches to this work, the epitome of many years of labour devoted to the study of the great central area of the North American continent." In commemorating Rydberg, Rendle referred to the posthumous paper "Taxonomic notes on the Flora of the Prairies and Plains of Central North America," (Rydberg, 1931) published in the new taxonomic journal of the Garden, *Brittonia.* This contained comments and notes which Rydberg would probably have reproduced in the preface to the flora had he lived a few weeks longer. Rendle remarked that Rydberg accepted many "segregations" apparently con amore and added that "Some of the new combinations will become invalid by obvious additions that are pending to the list of *nomina conservanda,*" under the new rules.

Two major reviews of the *Flora* were written by botanists not prejudiced against Rydberg's taxonomy: Rosendahl, Rydberg's companion on his last field trip, and Gleason. Rosendahl (1932) pointed out that the choice of the subject was altogether fitting: Rydberg's knowledge of the flora of the prairies and plains was unsurpassed. His botanical career started in the midst of this area and "To it he returned, with unabated interest and zeal, to carry on to the very end and bring forth what in the reviewer's opinion is much the best of his numerous systematic works." Rosendahl was gratified ". . . to find that a host of so-called species of doubtful validity have been relegated to well-deserved synonymy." This statement is remarkable because in the past Rydberg's severest critics had always said that he had done just the opposite. "In this the author has exercised unbiased judgment, for a considerable number of forms so treated are of his own former making." Rosendahl did find fault with dismemberment of large genera, preferring to keep them intact and to assign sectional rank to morphologically discordant groups of species. "But upon this question differences of opinion will always obtain and it is futile and out of place to pursue the discussion further here." Even if a taxonomist is not in full sympathy with Rydberg's principles ". . . he will frankly recognize that it is a work of outstanding merit" (Rosendahl, 1932).

Rydberg's younger colleague at the Garden, Henry Gleason, reviewed the book in *Torreya.* Gleason (1932) welcomed the flora as the first for the region, and expressed the opinion that diversity in genera and species required a hierarchy of categories, rather than a simple choice of genus and species. The review deals more with Gleason's own opinions than with the book itself, except for trivial details such as the number of pages of the glossary.

Apart from Rosendahl's review nothing impartial and substantial seems to have been written of this flora, which in practice was frequently used for at least half a century and was reprinted as late as 1965. Except for its archival interest it is now replaced by a modern flora (Great Plains Flora Association, 1986).

Rydberg's best friends were members of the Swedish Lutheran churches; at the Garden he was closest to Barnhart, bibliographer and, in general erudition and sympathy for his fellow botanists, outstanding in the group of "loners" at the Garden. Barnhart's obituary is precise in fact and gentle in approach. It concludes with a reference to Rydberg's second priority in life, religion, thereby adding a dimension to the picture of Rydberg as a man (Barnhart, 1931):

> *No account of Per Axel Rydberg would be complete without some reference to his religious life. Brought up in the state church of Sweden, sometimes described as the most strongly Lutheran of all nations, he was steadfast in his loyalty to Lutheranism as long as he lived. During the earlier years in America, he was a member of Swedish Lutheran churches affiliated with the Augustana Synod, and long after he came to the Bronx to live he worshipped regularly in a Swedish church in Mount Vernon, several miles from his home. As his children grew up, however, he felt that they should receive their religious instruction in the English language, and this was his chief motive in transferring his membership to Grace Lutheran Church of Bedford Park, where all the services are in English. It was in this church that his many friends gathered, on the afternoon of Monday, July 27, to honor his memory. There were three services: a private one, at the home, in Swedish; a public one, at the church, in English; and a brief commitment service, at the grave, in Woodlawn cemetery, where rest the mortal remains of one who, all through his life, was faithful to the service of his science and his God.*

Literature Cited

Barneby, R. C. 1964. Atlas of North American *Astragalus.* Mem. New York Bot. Gard. **13:** 1–1188.

Barnett, H. L. 1958. Ernst Athearn Bessey. Mycologia **50:** 1–5.

Barnhart, J. H. 1931. Per Axel Rydberg. J. New York Bot. Gard. **32:** 229–233.

Bessey, C. E. 1892. A second edition of Webber's "Appendix to the catalogue of the flora of Nebraska." Contr. Bot. Dept. Univ. Nebraska II. **3:** 1–55.

Britton, N. L. 1900. Report on exploration and collection during 1900. J. New York Bot. Gard. **1:** 183–184.

Coulter, J. M. 1885. Manual of the botany of the Rocky Mountain region. American Book Co., New York, NY. 453 pp.

———. 1915. Charles Edwin Bessey. Bot. Gaz. **60:** 72–73.

——— **& A. Nelson.** 1909. New manual of botany of the central Rocky Mountains. American Book Co., New York, NY. 646 pp.

Coville, F. V. 1901. [Review of] Britton's manual of the flora of the northern states and Canada. Torreya **1:** 133–134.

Cowles, H. C. 1908. An ecological aspect of the conception of species. Amer. Naturalist **42:** 265–271.

Cox, C. F. 1903. William Earl Dodge. J. New York Bot. Gard. **4:** 169–170.

Fernald, M. L. 1936. Dates of publication of Rydberg's Flora of the Rocky Mts. and adjacent plains. Rhodora **38:** 329–331.

Gleason, H. A. 1932. [Review of] Rydberg's flora of the prairies and plains. Torreya **32:** 130–133.

Great Plains Flora Association. 1986. Flora of the Great Plains. University Press of Kansas, Lawrence, Kansas. 392 pp.

Greene, E. L. 1910. Rocky Mountain botany, a general review. Amer. Midl. Naturalist **1:** 189–194.

Jones, M. E. 1910. [Review of] Flora of Colorado. Contr. W. Bot. **13:** 45.

———. 1929. [Review of] Rydberg's flora of the Rocky Mountains. Contr. W. Bot. **15:** 9–14.

———. 1934. P. A. Rydberg. Contr. W. Bot. **18:** 141–142.

Kiener, W. 1951. P. A. Rydberg's "Report of a botanical exploration, made in western Nebraska in the summer of 1891, as a special agent of the Department of Agriculture." Nebraska History **32:** 217–239, 306–331.

Lenz, L. W. 1986. Marcus E. Jones. Rancho Santa Ana Botanic Garden, Claremont, CA. 486 pp.

Léveillé, A. A. H. 1902. Monographie du genre *Onothera.* **1:** 1–138.

Merrill, E. D. 1938. Nathaniel Lord Britton 1859–1934. Natl. Acad. Sci. Biog. Mem. **19:** 147–160.

Nelson, A. 1899. New plants from Wyoming—X. Bull. Torrey Bot. Club **26:** 480–487.

———. 1900. [Review of] The flora of Montana. Bot. Gaz. **30:** 61–64.

———. 1901. New plants from Wyoming—XIII. Bull. Torrey Bot. Club **28:** 223–235.

———. 1906. [Review of] Rydberg's flora of Colorado. Torreya **6:** 237–239.

———. 1924. Taxonomic studies. Univ. Wyoming Publ. Sci., Bot. **1:** 47–68.

Rendle, A. B. 1933. [Review of] Flora of the prairies and plains of central north America. J. Bot. **71:** 53–54.

Rosendahl, C. O. 1932. [Review of] Rydberg's flora of the prairies and plains. J. New York Bot. Gard. **33:** 201–204.

Rydberg, P. A. 1882. *Cardamine pratensis* L. var. *acaulis*. Bot. Not. **1882:** 63.

———. 1895a. Rosales. Flora of Nebraska, pt **21:** 1–82. University of Nebraska, Lincoln, NB.

———. 1895b. Report of a botanical exploration of the sandhill region of central Nebraska made in the summer of 1893. M.A. Thesis, University of Nebraska, Lincoln, NB. 221 pp.

———. 1895c. Flora of the Sand Hills of Nebraska. Contr. U.S. Natl. Herb. **3:** 133–203.

———. 1896a. Notes on *Potentilla*—I. Bull. Torrey Bot. Club **23:** 244–248.

———. 1896b. Flora of the Black Hills of South Dakota. Contr. U.S. Natl. Herb. **3:** 463–536.

———. 1898. A monograph of the North American Potentilleae. Mem. Dept. Bot. Columbia Coll. **2:** 1–233.

———. 1900. Catalogue of the flora of Montana and the Yellowstone National Park. Mem. New York Bot. Gard. **1:** 1–492.

———. 1901a. Report of Dr. P. A. Rydberg, assistant curator, on the Scandinavian botanical gardens. J. New York Bot. Gard. **2:** 129–136.

———. 1901b. "When in Rome do as the Romans do." Torreya **1:** 61–65.

———. 1903. [Review of] A recent monograph of *Campanula rotundifolia* and its allies. Torreya **3:** 9–12.

———. 1905a [1906]. *Astragalus* and its segregates as represented in Colorado. Bull. Torrey Bot. Club **32:** 657–668.

———. 1905b. Explorations in Utah. J. New York Bot. Gard. **6:** 158–165.

———. 1905c. Penthoraceae. N. Amer. Fl. **22:** 75.

———. 1905d. Parnassiaceae. N. Amer. Fl. **22:** 77–80.

———. 1906. Flora of Colorado. Colorado Agric. Exp. Sta. Bull. **100:** 1–448.

———. 1911. Report of a trip to southeastern Utah in the summer of 1911. J. New York Bot. Gard. **12:** 237–253.

———. 1917 [1918]. Flora of the Rocky Mountains and adjacent plains. Published by the author, New York, NY. 1110 pp.

———. 1926a. Botanizing in the higher Alleghany Mountains—I. West Virginia. J. New York Bot. Gard. **27:** 1–6.

———. 1926b. Botanizing in the higher Alleghany Mountains II. West Virginia and Tennessee. J. New York Bot. Gard. **27:** 33–40.

———. 1929a. Fabaceae: Galegeae (pars). N. Amer. Fl. **24:** 251–314.

———. 1929b. Fabaceae: Galegeae (pars). N. Amer. Fl. **24:** 315–378.

———. 1929c. Fabaceae: Galegeae (pars). N. Amer. Fl. **24:** 379–462.

———. 1929d. Scylla or Charybdis. Proc. Internat. Cong. Pl. Sci. **2:** 1539–1551.

———. 1931. Taxonomic notes on the flora of the prairies and plains of central North America. Brittonia **1:** 79–104.

———. 1932. Flora of the prairies and plains of central North America. New York Botanical Garden, New York, NY. 969 pp.

Setchell, W. A. 1938. Marshall Avery Howe 1867–1936. Natl. Acad. Sci. Biog. Mem. 19: 243–258.

Small, J. K. 1896. [Review of] Flora of Nebraska, Part 21, Rosales. Per Axel Rydberg. Bull. Torrey Bot. Club **23:** 217.

———. 1900. The herbarium. J. New York Bot. Gard. **1:** 33–38.

——— **& P. A. Rydberg.** 1905a. Saxifragaceae. N. Amer. Fl. **22:** 81–158.

——— **&** ———. 1905b. Hydrangeaceae. N. Amer. Fl. **22:** 159–178.

Standley, P. C. 1918. [Review of] Rydberg's flora of the Rocky Mountains. Torreya **18:** 91–94.

Weber, W. A. 1965. Theodore Dru Alison Cockerell, 1866–1948. Univ. Colo. Stud., Ser. Bibl. **1:** 1–124.

Williams, R. L. 1984. Aven Nelson of Wyoming. Colorado Associated Univ. Press, Boulder, CO. 407 pp.

Witasek, J. 1902. Ein Beitrag zur Kenntnis der Gattung *Campanula*. Alfred Hölder, K.U.K. Hof-Und, Universitats-Buchandler, Wein. 106 pp.

A Rydberg Eponymy

Allium rydbergii Macbr., Contr. Gray Herb. **56:** 7. 1918.

Arnica rydbergii E. L. Greene, Pittonia **4:** 36. 1899.

Astragalus mortonii forma *rydbergii* Gandoger, Bull. Soc. Bot. France **48:** 16. 1902.

A. perianus Barneby, Mem. New York Bot. Gard. **13:** 972. 1964.

A. rydbergianus Tidestrom, Proc. Biol. Soc. Wash. **50:** 19. 1937.

A. rydbergii Macbr., Contr. Gray Herb. **65:** 37. 1922.

Atriplex rydbergii Standl., N. Amer. Fl. **21:** 47. 1916.

Cirsium rydbergii Petrak, Beih. Bot. Centralbl. **35:** 315. 1917.

Dalea rydbergii Tidestrom *in* Tidestrom & Kittell, Fl. Ariz. New Mex. 183. 1941.

Descurainia rydbergii O. E. Schulz, Pflanzenr. IV, **105:** 319. 1924.

Erigeron rydbergii Cronq., Brittonia **6:** 191. 1947.

Eriogonum rydbergii E. L. Greene, Pittonia **5:** 68. 1902.

Eupatorium rydbergii Britton, Man. fl. n. states 921. 1901.

Festuca ovina var. *rydbergii* St. Yves, Candollea **2:** 245. 1925.

Geum macrophyllum var. *rydbergii* Farwell, Pap. Michigan Acad. Sci. **23:** 129. 1938.

Haplopappus rydbergii S. F. Blake, Contr. U.S. Natl. Herb. **25:** 545. 1925.

Helianthus rydbergii Britton, Man. fl. n. states 993. 1901.

Heuchera rubescens var. *rydbergiana* Rosendahl, Butters & Lakela, Minn. Stud. Pl. Sci. **2:** 92. 1936.
Horkelia rydbergii Elmer, Bot. Gaz. **39:** 50. 1905.
Lupinus rydbergii Blankinship, Montana Agric. Coll. Sci. Stud., Bot. **1:** 78. 1905.
Oenothera rydbergii House, New York State Mus. Bull. **233-234:** 61. 1921.
Parrya rydbergii Botsch., Not. Syst. Inst. Bot. Komarov Acad. Sci. USSR **17:** 178. 1955.
Penstemon rydbergii A. Nels., Bull. Torrey Bot. Club **25:** 281. 1898.
Physalis rydbergii Fernald, Proc. Amer. Acad. Arts **35:** 569. 1900.
Puccinia rydbergii Garrett, Mycologia **6:** 251. 1914.
Quercus rydbergiana Cockerell, Torreya **3:** 7. 1903.
Rhus rydbergii Small *in* Rydberg, Mem. New York Bot. Gard. **1:** 268. 1900.
Rydbergia E. L. Greene, Pittonia **3:** 270. 1898.
Rydbergiella Fedde & Sydow, Bot. Jahresber. **33:** 534. 1906.
Salix rydbergii A. A. Heller, Catalogue N. Amer. Plants, ed 2, 4. 1900.
Saxifraga rydbergii Small *in* Rydberg, Mem. New York Bot. Gard. **1:** 194. 1900.
Senecio rydbergii A. Nels. *in* Coulter & Nelson, New man. bot. Rocky Mt. 582. 1909.
Sphaeralcea rydbergii Tidestrom, Contr. U.S. Natl. Herb. **25:** 356. 1925.
Trifolium rydbergii E. L. Greene, Pittonia **3:** 222. 1897.
Verbena rydbergii Moldenke, Revista Sudamer. Bot. **4:** 19. 1937.
Viola rydbergii E. L. Greene, Pittonia **5:** 27. 1902.

Bibliography of P. A. Rydberg

1882

Cardamine pratensis L. var. *acaulis*. Bot. Not. **1882:** 63.

1891

The flora of the high Nebraska plains. Amer. Naturalist **25:** 485, 486 (May).

1892

Some grasses of southwestern Nebraska. Pages 134–137 *in* C. E. Bessey. Sixth annual report of the botanist. Nebraska State Board of Agric. Annual Rep. **1892:** 122–137. Lincoln, Nebraska.

1893

On the American Black Cottonwood. Bull. Torrey Bot. Club **20:** 46–50 (15 Feb).

1894

A revision of the nomenclature of the Nebraska Polypetalae. Bot. Surv. Nebraska **3:** 21–39. University of Nebraska. (18 Jun).

The grasses of central Nebraska. Pages 69–73 *in* C. E. Bessey. Eighth annual report of the botanist. Nebraska State Board of Agric. Annual Rep. **1894:** 65–129. Lincoln, NB.

1895

New species of *Physalis*. Bull. Torrey Bot. Club **22:** 306–308 (31 Jul).

Flora of the Sand Hills of Nebraska. Contr. U.S. Natl. Herb. **3:** 133–203 (14 Sep).

Rosales. Flora of Nebraska, pt. **21:** 1–82. University of Nebraska, Lincoln, Nebraska (30 Dec).

Report of a botanical exploration of the sandhill region of central Nebraska made in the summer of 1893. M.A. Thesis, University of Nebraska, Lincoln, Nebraska. 221 pp.

1896

Notes on *Potentilla*—I. Bull. Torrey Bot. Club **23:** 244–248 (30 Jun).

Notes on *Potentilla*—II. Bull. Torrey Bot. Club **23:** 259–265 (20 Jul).

Notes on *Potentilla*—III. Bull. Torrey Bot. Club **23:** 301–306 (20 Aug).

Notes on *Potentilla*—IV. Bull. Torrey Bot. Club **23:** 394–399 (25 Oct).

Notes on *Potentilla*—V. Bull. Torrey Bot. Club **23:** 429–435 (25 Nov).

Flora of the Black Hills of South Dakota. Contr. U.S. Natl. Herb. **3:** 463–536 (13 Jun).

The North American species of *Physalis* and related genera. Mem. Torrey Bot. Club **4:** 297–374 (15 Sep).

1897

Notes on *Potentilla*—VI. Bull. Torrey Bot. Club **24:** 1–13 (28 Jan).

Notes on two western plants. Bull. Torrey Bot. Club **24:** 90–92 (28 Feb).

Rarities from Montana—I. Bull. Torrey Bot. Club **24:** 188–192 (24 Apr).

Rarities from Montana—II. Bull. Torrey Bot. Club **24:** 243–253 (29 May).

Rarities from Montana—III. Bull. Torrey Bot. Club **24:** 292–299 (29 Jun).

Antennaria dioica and its North American allies. Bull. Torrey Bot. Club **24:** 299–304 (29 Jun).

Potentilla. Pages 208–216 *in* N. L. Britton & A. Brown, Illustrated Flora of the Northern United States, Canada and the British possessions **2:** 1–643. Charles Scribner's Sons, New York, NY (31 May).

A report upon the grasses and forage plants of the Rocky Mountain Region. U.S.D.A. Div. of Agrostol. Bull. **5:** 1–48 (19 Feb) (senior author with C. L. Shear).

1898

Some changes in the nomenclature of North American Rosaceae. Bull. Torrey Bot. Club **25:** 54–56 (25 Jan).

A monograph of the North American Potentilleae. Mem. Dept. Bot. Columbia Coll. **2:** 1–233 (25 Nov).

Physalis. Pages 125–132 *in* N. L. Britton & A. Brown, Illustrated Flora of the Northern United States, Canada and the British possessions **3:** 1–588. Charles Scribner's Sons, New York, NY (20 Jun).

Horkelia, Ivesia. Pages 179–182 *in* T. J. Howell, A Flora of Northwest America **2:** 113–274. Published by the author, Portland, OR (1 Apr).

1899

What is *Prunus insititia*? Bot. Gaz. **28:** 423–425 (10 Jan 1900).

Caespitose willows of arctic America and the Rocky Mountains. Bull. New York Bot. Gard. **1:** 257–278 (13 Apr).

New species from the western United States. Bull. Torrey Bot. Club **26:** 541–546 (16 Oct).

Delphinium carolinianum and related species. Bull. Torrey Bot. Club **26:** 582–587 (15 Nov).

1900

Studies on the Rocky Mt. Flora—I. Bull. Torrey Bot. Club **27:** 169–189 (21 Apr).

What is *Prunus insititia*? Bull. Torrey Bot. Club **27:** 195, 196 (21 Apr).

Studies on the Rocky Mt. Flora—II. Bull. Torrey Bot. Club **27:** 528–538 (26 Oct).

Studies on the Rocky Mt. Flora—III. Bull. Torrey Bot. Club **27:** 614–636 (29 Dec).

Catalogue of the flora of Montana and the Yellowstone National Park. Mem. New York Bot. Gard. **1:** 1–492 (15 Feb).

Plant geography of North America. Science II, **12:** 870–873 (7 Dec).

1901

Contributions to the botany of the Yukon Territory 4. An enumeration of the flowering plants collected by R. S. Williams and by J. B. Tarleton. Bull. New York Bot. Gard. **2:** 149–187 (27 May) (junior author with N. L. Britton).

The oaks of the Continental Divide north of Mexico. Bull. New York Bot. Gard. **2:** 187–233 (27 May).

Studies on the Rocky Mt. Flora—IV. Bull. Torrey Bot. Club **28:** 20–38 (31 Jan).

Further studies on the Potentilleae. Bull. Torrey Bot. Club **28:** 173–183 (27 Mar).

Studies on the Rocky Mt. Flora—V. Bull. Torrey Bot. Club **28:** 266–284 (21 May).

Studies on the Rocky Mt. Flora—VI. Bull. Torrey Bot. Club **28:** 499–513 (30 Sep).

The American species of *Limnorchis* and *Piperia,* north of Mexico. Bull. Torrey Bot. Club **28:** 605–643 (25 Nov).

Report of Dr. P. A. Rydberg, assistant curator, on the Scandinavian Botanical Gardens. J. New York Bot. Gard. **2:** 129–136 (Sep).

Economy in Nature. Torreya **1:** 10 (29 Jan).

The North American twinflowers. Torreya **1:** 52–54 (22 May).

"When in Rome do as the Romans do." Torreya **1:** 61–65 (20 Jun).

Orchidaceae. Pages 289–306; Salicaceae. Pages 307–320; Ranunculaceae. Pages 411–432; Saxifragaceae. Pages 476–484; Rosaceae. Pages 490–514; *Physalis.* Pages 809–812 *in* N. L. Britton, Manual of the flora of the Northern States and Canada, Henry Holt & Co., New York, NY. 1043 pp. (24 Aug).

1902

Studies on the Rocky Mt. Flora—VII. Bull. Torrey Bot. Club **29:** 145–160 (24 Mar).

Studies on the Rocky Mt. Flora—VIII. Bull. Torrey Bot. Club **29:** 232–246 (24 Apr).

Studies on the Rocky Mt. Flora—IX. Bull. Torrey Bot. Club **29:** 680–693 (30 Dec).

Our yellow lady's slippers. Torreya **2:** 84–87 (12 Jun).

A new station for *Isotria affinis.* Torreya **2:** 143 (30 Aug).

Is the white-fruited strawberry of Pennsylvania a native species? Torreya **2:** 158, 159 (30 Sep).

1903

Studies on the Rocky Mt. Flora—X. Bull. Torrey Bot. Club **30:** 247–262 (24 Apr).

Some generic segregations. Bull. Torrey Bot. Club **30:** 271–281 (16 May).

[Review of] A recent monograph of *Campanula rotundifolia* and its allies. Torreya **3:** 9–12 (26 Jan).

Key to the orders. Pages vii–xii; *Rubus.* Pages 515–519; *Geoprumnon, Astragalus, Hamosa, Holcophacos, Tium, Phaca, Xylophacos.* Pages 615–620; *Physalis.* Pages 981–987 *in* J. K. Small, Flora of the Southeastern United States. Published by the author, New York, NY. 1370 pp. (22 Jul).

1904

Studies on the Rocky Mt. Flora—XI. Bull. Torrey Bot. Club **31:** 399–410 (29 Jul).

Studies on the Rocky Mt. Flora—XII. Bull. Torrey Bot. Club **31:** 555–575 (1 Nov).

Studies on the Rocky Mountain Flora—XIII. Bull. Torrey Bot. Club **31:** 631–655 (9 Jan 1905).

[Review of] Howell's *Flora of Northwest America.* Torreya **4:** 70–72 (13 May).

1905

Studies on the Rocky Mt. Flora—XIV. Bull. Torrey Bot. Club **32:** 123–138 (19 Apr).

Studies on the Rocky Mt. Flora—XV. Bull. Torrey Bot. Club **32:** 597–610 (6 Dec).

Astragalus and its segregates as represented in Colorado. Bull. Torrey Bot. Club **32:** 657–668 (22 Jan 1906).

Explorations in Utah. J. New York Bot. Gard. **6:** 158–165 (Oct).

Penthoraceae. N. Amer. Fl. **22:** 75 (22 May).

Parnassiaceae. N. Amer. Fl. **22:** 77–80 (22 May).

Saxifragaceae. N. Amer. Fl. **22:** 81–158 (18 Dec) (junior author with J. K. Small).
Hydrangeaceae. N. Amer. Fl. **22:** 159–178 (18 Dec) (junior author with J. K. Small).

1906

Studies on the Rocky Mt. Flora—XVI. Bull. Torrey Bot. Club **33:** 137–161 (7 Apr).
Flora of Colorado. Colorado State Agric. Exp. Sta. Bull. **100:** 1–448 (Mar).
Grayia or *Eremosemium.* Torreya **6:** 8–10 (25 Jan).
Bossekia or *Rubacer.* Torreya **6:** 165–169 (25 Aug).

1907

Scandinavians who have contributed to the knowledge of the flora of North America. Augustana Libr. Publ. **6:** 1–49. Augustana College, Rock Island, IL.
Studies on the Rocky Mt. Flora—XVII. Bull. Torrey Bot. Club **34:** 35–50 (27 Feb).
Studies on the Rocky Mt. Flora—XVIII. Bull. Torrey Bot. Club **34:** 417–437 (10 Oct).
Linnaeus and American botany. J. New York Bot. Gard. **8:** 124–135 (Jun).
Linnaeus and American botany. Science II, **26:** 65–71 (19 Jul).
The genus *Pilosella* in North America. Torreya **7:** 157–162 (21 Aug).

1908

Notes on *Philotria* Raf. Bull. Torrey Bot. Club **35:** 457–465 (29 Sep).
Notes on Rosaceae—I. Bull. Torrey Bot. Club **35:** 535–542 (30 Nov).
Rosaceae (pars). N. Amer. Fl. **22:** 239–292 (12 Jun).
Rosaceae (pars). N. Amer. Fl. **22:** 293–388 (20 Nov).

1909

Notes on Rosaceae—II. Bull. Torrey Bot. Club **36:** 397–407 (14 Aug).
Studies on the Rocky Mt. Flora—XIX. Bull. Torrey Bot. Club **36:** 531–541 (1 Oct).
Studies on the Rocky Mt. Flora—XX. Bull. Torrey Bot. Club **36:** 675–698 (28 Dec).
The flowers and fruit of the turtle-grass. J. New York Bot. Gard. **10:** 261–264 (Nov).
Sparganiaceae. N. Amer. Fl. **17:** 5–10 (30 Jun).
Elodeaceae. N. Amer. Fl. **17:** 67–71 (30 Jun).
Hydrocharitaceae. N. Amer. Fl. **17:** 73, 74 (30 Jun).

1910

Studies on the Rocky Mt. Flora—XXI. Bull. Torrey Bot. Club **37:** 127–148 (31 Mar).
Studies on the Rocky Mt. Flora—XXII. Bull. Torrey Bot. Club **37:** 313–335 (21 Jul).
Notes on Rosaceae—III. Bull. Torrey Bot. Club **37:** 375–386 (29 Jul).
Studies on the Rocky Mt. Flora—XXIII. Bull. Torrey Bot. Club **37:** 443–471 (5 Oct).
Notes on Rosaceae—IV. Bull. Torrey Bot. Club **37:** 487–502 (28 Oct).
Studies on the Rocky Mt. Flora—XXIV. Bull. Torrey Bot. Club **37:** 541–557 (30 Nov).
Balsaminaceae. N. Amer. Fl. **25:** 93–96 (3 Jun).
Limnanthaceae. N. Amer. Fl. **25:** 97–100 (3 Jun).
Zygophyllaceae. N. Amer. Fl. **25:** 103–116 (3 Jun) (junior author with A. M. Vail).

1911

Studies on the Rocky Mt. Flora—XXV. Bull. Torrey Bot. Club **38:** 11–23 (15 Feb).
Notes on Rosaceae—V. Bull. Torrey Bot. Club **38:** 79–89 (7 Mar).
Notes on Rosaceae—VI. Bull. Torrey Bot. Club **38:** 351–367 (21 Aug).
Report of a trip to southeastern Utah in the summer of 1911. J. New York Bot. Gard. **12:** 237–253 (Nov).
List of plants collected on the Peary Arctic Expedition of 1905–06 and 1908–09 with a general description of the flora of northern Greenland and Ellesmere Land. Torreya **11:** 249–259 (20 Dec).

1912

Studies on the Rocky Mt. Flora—XXVI. Bull. Torrey Bot. Club **39:** 99–111 (18 Apr).
Studies on the Rocky Mt. Flora—XXVII. Bull. Torrey Bot. Club **39:** 301–328 (23 Jul).
List of plants collected on the Peary Arctic Expedition (continued). Torreya **12:** 1–11 (22 Jan).
Phytogeography and its relation to taxonomy and other branches of science. Torreya **12:** 73–85 (17 Apr).
Epipactis vs. *Peramium.* Torreya **12:** 89, 90 (17 Apr).

1913

Studies on the Rocky Mt. Flora—XXVIII. Bull. Torrey Bot. Club **40:** 43–74 (28 Mar).
Studies on the Rocky Mt. Flora—XXIX. Bull. Torrey Bot. Club **40:** 461–485 (10 Sep).
Phytogeographical notes on the Rocky Mountain Region I. Alpine region. Bull. Torrey Bot. Club **40:** 677–686 (6 Jan 1914).
Rosaceae (pars). N. Amer. Fl. **22:** 389–480 (23 Dec).

1914

Phytogeographical notes on the Rocky Mountain region II. Origin of the alpine flora. Bull. Torrey Bot. Club **41:** 89–103 (23 Mar).
Notes on Rosaceae—VII. Bull. Torrey Bot. Club **41:** 319–332 (22 Jul).
Phytogeographical notes on the Rocky Mountain region III. Formations in the alpine flora. Bull. Torrey Bot. Club **41:** 459–474 (8 Oct).
Notes on Rosaceae—VIII. Bull. Torrey Bot. Club **41:** 483–503 (28 Oct).
Lennoaceae. N. Amer. Fl. **29:** 19, 20 (31 Aug).
Pyrolaceae. N. Amer. Fl. **29:** 21–33 (31 Aug).
Carduaceae: Helenieae. N. Amer. Fl. **34:** 1–75 (31 Dec).
List of plants collected on the Stefanson-Anderson Arctic Expedition, 1908–12. Torreya **14:** 65, 66 (8 Apr).

1915

Phytogeographical notes on the Rocky Mountain region IV. Forests of the subalpine and montane zones. Bull. Torrey Bot. Club **42:** 11–25 (29 Jan).

Notes on Rosaceae—IX. Bull. Torrey Bot. Club **42:** 117–160 (16 Mar).

Notes on Rosaceae—X. Bull. Torrey Bot. Club **42:** 463–479 (22 Sep).

Phytogeographical notes on the Rocky Mountain region V. Grasslands of the subalpine and montane zones. Bull. Torrey Bot. Club **42:** 629–642 (11 Dec).

Carduaceae: Helenieae, Tageteae. N. Amer. Fl. **34:** 81–180 (28 Jul).

1916

Phytogeographical notes on the Rocky Mountain region VI. Distribution of the subalpine plants. Bull. Torrey Bot. Club **43:** 343–364 (14 Aug).

Vegetative life zones of the Rocky Mountain region. Mem. New York Bot. Gard. **6:** 477–499 (31 Aug).

Carduaceae: Tageteae, Anthemideae. N. Amer. Fl. **34:** 181–288 (29 Dec).

[Review of] Piper and Beattie's *Flora of the Northwest Coast.* Torreya **16:** 143–146 (15 Jun).

1917

Notes on Rosaceae—XI. Bull. Torrey Bot. Club **44:** 65–84 (27 Feb).

Phytogeographic notes on the Rocky Mountain region VII: Formations in the subalpine zone. Bull. Torrey Bot. Club **44:** 431–454 (15 Sep).

Flora of the Rocky Mountains and adjacent plains. Published by the author, New York, NY. 1110 pp. (early 1918, see Fernald, 1936, for a discussion of the date of publication).

1918

Rosaceae (conclusion). N. Amer. Fl. **22:** 481–533 (30 Dec).

Penthoraceae (additions & corrections). N. Amer. Fl. **22:** 548 (30 Dec).

Parnassiaceae (additions & corrections). N. Amer. Fl. **22:** 548, 549 (30 Dec).

Saxifragaceae (additions & corrections). N. Amer. Fl. **22:** 549–555 (30 Dec) (junior author with J. K. Small).

Hydrangeaceae (additions & corrections). N. Amer. Fl. **22:** 555–557 (30 Dec) (junior author with J. K. Small).

1919

Phytogeographical notes on the Rocky Mountain region VIII. Distribution of the montane plants. Bull. Torrey Bot. Club **46:** 295–327 (27 Sep).

A genus of plants intermediate between *Petalostemon* and *Parosela.* J. New York Bot. Gard. **20:** 64–66 (Mar).

Fabaceae: Psoraleae. N. Amer. Fl. **24:** 1–34, 40–64 (25 Apr).

Key to the Rocky Mountain Flora. 305 pp. Published by the author, New York, NY (Mar).

1920

Notes on Rosaceae—XII. Bull. Torrey Bot. Club **47:** 45–66 (10 Mar).

Phytogeographical notes on the Rocky Mountain region IX. Wooded formations of the montane zone of the southern Rockies. Bull. Torrey Bot. Club **47:** 441–454 (29 Oct).

Bembicidium. Page 68 *in* N. L. Britton, Descriptions of Cuban plants new to science. Mem. Torrey Bot. Club **16:** 57–118 (10 Sep).

Fabaceae: Psoraleae. N. Amer. Fl. **24:** 65–136 (22 Jan).

[Review of] Henry and Flood's *The Douglas Fir.* Torreya **20:** 102–104 (8 Jul).

1921

Two new species from Arizona. Amer. Bot. (Binghamton) **27:** 61–63 (May).

Notes on Rosaceae—XIII. Bull. Torrey Bot. Club **48:** 159–172 (1 Aug).

Phytogeographical notes on the Rocky Mountain region X. Grasslands and other open formations of the montane zone of the southern Rockies. Bull. Torrey Bot. Club **48:** 315–326 (28 Feb 1922).

1922

Ambrosiaceae. N. Amer. Fl. **33:** 3–44 (15 Sep).

Carduaceae. N. Amer. Fl. **33:** 45, 46 (15 Sep).

Flora of the Rocky Mountains and adjacent plains. Ed 2. Published by the author, New York, NY. 1144 pp. (21 Jan 1923, see Fernald, 1936, for a discussion of the date of publication).

1923

Ribes cereum. Addisonia **8:** 7, 8 (15 May).

Rosa palustris. Addisonia **8:** 37, 38 (10 Nov).

Salvia pitcheri. Addisonia **8:** 53, 54 (5 Feb 1924).

Genera of North American Fabaceae I. Tribe Galegeae. Amer. J. Bot. **10:** 485–498 (27 Nov).

Notes on Rosaceae—XIV. Bull. Torrey Bot. Club **50:** 61–71 (23 Feb).

Notes on Fabaceae—I. Bull. Torrey Bot. Club **50:** 177–187 (25 May).

Notes on Fabaceae—II. Bull. Torrey Bot. Club **50:** 261–272 (17 Aug).

A new genus of senecioid composites. J. Wash. Acad. Sci. **13:** 287–289 (19 Jul).

Fabaceae: Indigofereae, Galegeae (pars). N. Amer. Fl. **24:** 137–200 (16 Jul).

1924

Erythronium grandiflorum. Addisonia **9:** 23–24 (30 Jul).

Genera of North American Fabaceae II. Tribe Galegeae (continued). Amer. J. Bot. **11:** 470–482 (14 Aug).

Notes on Fabaceae—III. Bull. Torrey Bot. Club **51:** 13–23 (8 Feb).

The section *Tuberarium* of the genus *Solanum* in Mexico and Central America. Bull. Torrey Bot. Club **51:** 145–154 (17 May).

The section *Tuberarium* of the genus *Solanum* in Mex-

ico and Central America. Concluding paper. Bull. Torrey Bot. Club **51:** 167–176 (13 Jun).
Some Senecioid genera—I. Bull. Torrey Bot. Club **51:** 369–378 (18 Sep).
Some Senecioid genera—II. Bull. Torrey Bot. Club **51:** 409–420 (24 Oct).
Plants used by ancient American Indians. J. New York Bot. Gard. **25:** 204, 205 (Jul).
Fabaceae: Galegeae (pars). N. Amer. Fl. **24:** 201–250 (4 Oct).
Some extinct or lost and rediscovered plants—I. Torreya **24:** 98–101 (31 Dec).

1925

Notes on Fabaceae—IV. Bull. Torrey Bot. Club **52:** 143–156 (18 May).
Notes on Fabaceae—V. Bull. Torrey Bot. Club **52:** 229–235 (7 Jun).
Notes on Fabaceae—VI. Bull. Torrey Bot. Club **52:** 365–372 (20 Nov).
Alpine flowers of the Rocky Mountains. J. New York Bot. Gard. **26:** 87, 88 (Apr).
Some extinct or lost and rediscovered plants—II. Torreya **25:** 96–98 (27 Oct).

1926

Gaillardia amblyodon. Addisonia **11:** 31, 32 (13 Oct).
Notes on Fabaceae—VII. Bull. Torrey Bot. Club **53:** 161–169 (1 Apr).
Botanizing in the higher Alleghany Mountains—I. West Virginia. J. New York Bot. Gard. **27:** 1–6 (Jan).
Botanizing in the higher Alleghany Mountains—II. West Virginia and Tennessee. J. New York Bot. Gard. **27:** 33–40 (Feb).
Botanizing in the higher Alleghany Mountains—III. North Carolina. J. New York Bot. Gard. **27:** 80–88 (Apr).
Two new species from the mountains of West Virginia. Torreya **26:** 29–33 (24 Apr).
Some extinct or lost and rediscovered plants—III. Torreya **26:** 86–89 (30 Oct).

1927

Notes on Fabaceae—VIII. Bull. Torrey Bot. Club **54:** 13–23 (22 Feb).
Notes on Fabaceae—IX. Bull. Torrey Bot. Club **54:** 321–336 (16 May).
Two native bigeneric hybrids. J. New York Bot. Gard. **28:** 227–228 (Sep).
Carduaceae: Liabeae, Neurolaeneae, Senecioneae (pars). N. Amer. Fl. **34:** 289–360 (22 Jun).
New species from the Blue Ridge. Torreya **27:** 84–90 (4 Nov).

1928

Sedum ternatum. Addisonia **13:** 21, 22 (11 Jul).
Genera of North American Fabaceae III. Tribe Psoraleae. Amer. J. Bot. **15:** 195–203 (27 Mar).
Genera of North American Fabaceae IV. Tribe Psoraleae (continued). Amer. J. Bot. **15:** 425–432 (8 Aug).
Genera of North American Fabaceae V. *Astragalus* and related genera. Amer. J.Bot. **15:** 584–595 (11 Jan 1929).
Notes on Fabaceae—X. Bull. Torrey Bot. Club **55:** 119–132 (15 Mar).
Notes on Fabaceae—XI. Bull. Torrey Bot. Club **55:** 155–164 (7 Apr).
Life zones of the Rocky Mountains. J. New York Bot. Gard. **29:** 282–284 (Nov).

1929

Bicuculla eximia. Addisonia **14:** 53, 54 (28 Feb 1930).
Genera of North American Fabaceae VI. *Astragalus* and related genera (continued). Amer. J. Bot. **16:** 197–206 (9 May).
Notes on Fabaceae—XII. Bull. Torrey Bot. Club **56:** 539–554 (17 Feb 1930).
Fabaceae: Galegeae (pars). N. Amer. Fl. **24:** 251–314 (19 Feb).
Fabaceae: Galegeae (pars). N. Amer. Fl. **24:** 315–378 (15 May).
Fabaceae: Galegeae (pars). N. Amer. Fl. **24:** 379–462 (26 Jul).
Scylla or Charybdis. Proc. Internat. Cong. Pl. Sci. **2:** 1539–1551.

1930

Genera of North American Fabaceae VII. *Astragalus* and related genera (continued). Amer. J. Bot. **17:** 231–238 (20 Mar).
Notes on Fabaceae—XIII. Bull. Torrey Bot. Club **57:** 397–407 (20 Jun 1931).

1931

Zizia bebbii. Addisonia **16:** 23, 24 (21 Jul).
A short phytogeography of the prairies and great plains of central North America. Brittonia **1:** 57–66 (21 Feb).
Taxonomic notes on the flora of the prairies and plains of central North America. Brittonia **1:** 79–104 (15 Oct).

1932

Flora of the prairies and plains of central North America. New York Botanical Garden, New York, NY. 969 pp. (Mar).
Portulacaceae. N. Amer. Fl. **21:** 279, 280, 290–328 (29 Dec).

1933

Physalis. Pages 1107–1112; *Acanthoxanthium, Xanthium.* Pages 1301–1303 *in* J. K. Small, Manual of the Southeastern Flora. Published by the author, New York, NY. 1554 pp. (30 Nov).

New Taxa Described by Per Axel Rydberg

In the following list of new taxa I have tried to establish type collections as well as the place of deposition of the holotypes. In attempting this I have encountered problems which I have not been able to solve. Nearly all these problems involve taxa described before 1900. It appears that once Rydberg was on the staff of the New York Botanical Garden the type concept, perhaps with Britton's influence, became ingrained in his works as he usually listed types as well as place of deposition for his new taxa.

The problems encountered are of two sorts. In some instances he listed a number of collections, now called syntypes, not designating any one as holotype. Some of these are resolved in that he wrote "TYPE" on the sheets (Fig. 3) and in these cases I have used the phrase—"designated by Rydberg in herb." Many of the rest of these await lectotypification by future monographers. The second problem is trying to locate where these specimens are deposited. To automatically assume that they are at NY is a trap which could lead to false conclusions. Rydberg saw the collections at GH, US, and Columbia University (now at NY) as early as the summer of 1894 and in his dissertation, published in 1898, he makes the following statement:

> *The study comprised most of the more important collections of this country, as those of Columbia University, the United States National Herbarium, Harvard University, Missouri Botanical Garden, University of Minnesota, California Academy of Sciences, Geological Survey of Canada, Iowa Agricultural College (including the Parry Collection), Lafayette College, New York Botanical Garden, College of Pharmacy of New York City, and the University of Nebraska.*
>
> *I hereby extend my thanks to the directors and curators of these institutions, as well as to the following persons who have kindly loaned me their private collections or sent specimens: S. B. Parish, San Bernardino, Cal.; Thomas Howell, Clackamas, Ore.; C. F. Baker, Auburn, Ala.; Professor Aven Nelson, University of Wyoming; Geo. E. Osterhout, New Windsor, Colo.; O. A. Farwell, Detroit, Mich., and others.*

Thus typification and deposition need to be handled case by case and conclusions from one case may not apply to the next. It is hoped that this information will make it easier for future workers to adequately solve these nomenclatural problems. Rydberg's nomina nova are not included.

The new taxa described in the *Flora of Colorado* have been considered by at least one author to be invalid (Weber, W. A., Phytologia **33:** 105, 106. 1976). An account of the problems and reasons for these inadvertently described new taxa is presented elsewhere (see Tiehm, Brittonia **41:** 152–155. 1989).

Data are from the specimen labels unless contained in brackets [ ]. If a country is not named, U.S.A. is understood.

The following abbreviations are used: BTBC = Bull. Torrey Bot. Club, NAF = N. Amer. Fl., HT = HOLOTYPE, IT = ISOTYPE, LT = LECTOTYPE, ILT = ISOLECTOTYPE, CT = CLASTOTYPE (KLEPTOTYPE), ST = SYNTYPE, T or Ts = GENERITYPE (s).

ACERACEAE

Negundo orizabense BTBC **40:** 55. 1913. **MEXICO. [Veracruz]:** Orizaba, 1853, *Muller s.n.* (HT: NY!—designated by Rydberg in herb.).

ANACARDIACEAE

Rhus rydbergii Small ex Rydb., Mem. New York Bot. Gard. **1:** 268. 1900. **Montana:** Great Falls, 5 Aug 1885, *Williams 291* (LT: NY!—Gillis, Rhodora **73:** 388. 1971; ILT: MONT).

APIACEAE

Aletes obovata BTBC **31:** 573. 1904. **Colorado:** Near Morrison, 27 Jul 1889, *Greene s.n.* (HT: NY!).

Angelica piperi Fl. Rocky Mts. 631, 1064. 1917 [1918]. **Washington. Walla Walla Co.:** Blue Mts., 15 Jul 1896, *Piper 2336* (HT: NY!; IT: US! as Columbia Co.).

Conioselinum pumilum Rose ex Rydb., Torreya **12:** 9. 1912. **CANADA. Labrador:** Caribou Island, Battle Harbor, 15 Aug 1909, *Goodsell 81* (HT: NY!; CT: US!).

Cynomarathrum latilobum BTBC **40:** 73. 1913. **Utah. Grand Co.:** Near Wilson Mesa, 1 Jul 1911, *Rydberg & Garrett 8371* (HT: NY!; IT: US!, UT!).

Glycosma maxima BTBC **40:** 67. 1913. **Utah:** Mt. Nebo, 15 Aug 1905, *Rydberg & Carlton 7585* (HT: NY!; IT: US).

Ligusticum brevilobum Fl. Rocky Mts. 613, 1064. 1917 [1918]. **Utah:** Aquarius Plateau, 5 Aug 1905, *Rydberg & Carlton 7473* (HT: NY!).

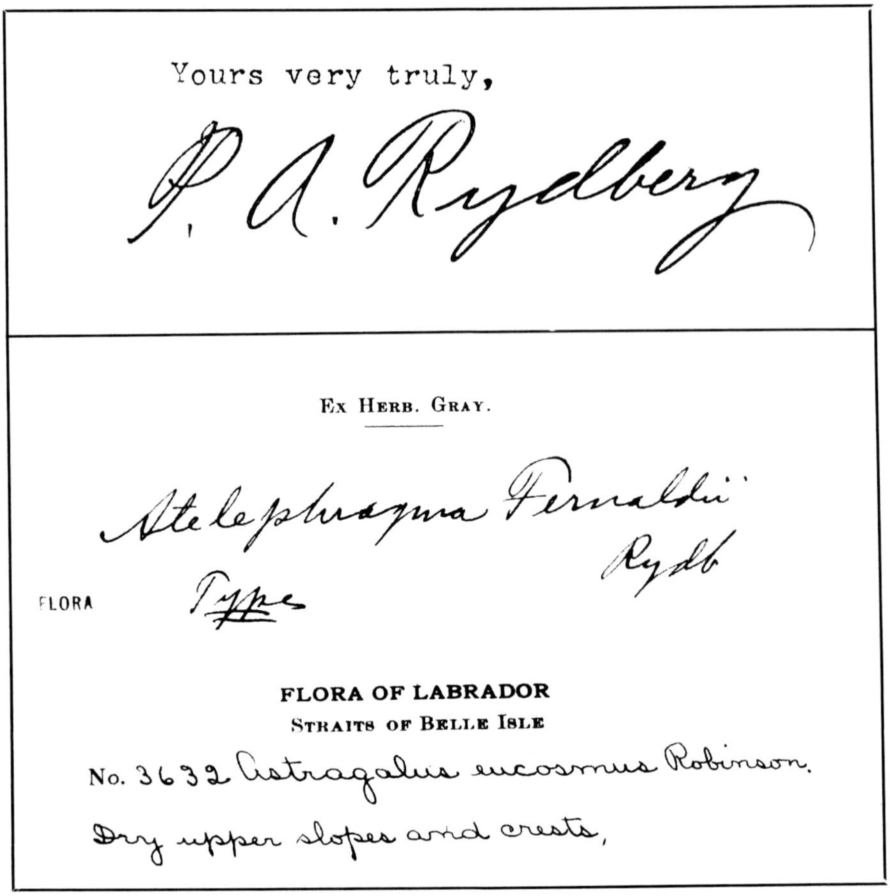

Fig. 3. [Upper] Signature taken from letter to M. A. Day dated 26 Dec 1908. Courtesy of Gray Herbarium Archives. [Lower] Characteristic herbarium annotation showing how Rydberg designated type specimens.

Musineon vaginatum Mem. New York Bot. Gard. **1:** 288. 1900. MONTANA: Bridger Mts., 15 Jun 1897, *Rydberg & Bessey 4626* (HT: NY! on 2 sheets; IT: MONT, NEB, RM!).

Phellopterus camporum BTBC **31:** 574. 1904. COLORADO: Mesas near Pueblo, 14 May 1900, *Rydberg & Vreeland 5825* (HT: NY!; IT: US!).

Pseudocymopterus aletifolius BTBC **31:** 574. 1904. COLORADO: Minnehaha, 16–22 Jun 1901, *Clements & Clements 94* (HT: NY!; IT: NEB, RM!).

P. montanus var. *multifidus* BTBC **31:** 574. 1904. NEW MEXICO: Range between Sapello & Pecos Rivers, 1–4 Aug 1900, *Cockerell & Cockerell s.n.* (LT: NY!—Wooton & Standley, Contr. U.S. Natl. Herb. **19:** 483. 1915).

P. versicolor Fl. Rocky Mts. 623, 1064. 1917 [1918]. UTAH: Aquarius Plateau, head of Poison Creek, 4 Aug 1905, *Rydberg & Carlton 7426* (HT: NY!).

Pseudopteryxia BTBC **40:** 71. 1913. T: *Cymopterus anisatus* A. Gray.

P. longiloba BTBC **40:** 72. 1913. UTAH: Abajo Mts., 17 Aug 1911, *Rydberg & Garrett 9761* (HT: NY!; IT: UT!).

Pseudoreoxis BTBC **40:** 73. 1913. Ts: *Cymopterus bipinnatus* S. Wats., *C. nivalis* S. Wats.

Washingtonia intermedia Mem. New York Bot. Gard. **1:** 289. 1900. MONTANA: Bridger Mts., 17 Jun 1897, *Rydberg & Bessey 4595* (HT: NY!; IT: NEB, US!).

APOCYNACEAE

Amsonia eastwoodiana BTBC **40:** 465. 1913. UTAH: Moab and vicinity, 1–2 Jul 1911, *Rydberg & Garrett 8468* (HT: NY! on 2 sheets; IT: RM!, US!, UT!).

Apocynum macranthum Fl. Rocky Mts. 669, 1065. 1917 [1918]. IDAHO. **Nez Perces Co.:** Big Potlatch River, 9 Jul 1892, *Sandberg, MacDougal & Heller 372* (HT: NY!).

A. scopulorum E. L. Greene ex Rydb., Fl. Colorado

269. 1906. COLORADO: Pagosa Springs, 17 Jul 1899, *Baker 527* (HT: NY!; IT: RM!, UC!, US!).

ASTERACEAE

Acanthambrosia NAF **33:** 22. 1922. T: *Franseria bryantii* Curran.

Achillea angustissima NAF **34:** 225. 1916. **OREGON:** Nasqually, *Wilkes Expedition 41* (HT: NY!).

A. fusca NAF **34:** 221. 1916. **WASHINGTON:** Mt. Paddo, 28 Sep 1893, *Suksdorf 1606* (HT: NY!).

A. lanulosa var. *alpicola* Mem. New York Bot. Gard. **1:** 426. 1900. **WYOMING. Uinta Co.:** Teton Forest Reserve, To-Gwa-Tee Pass, Aug 1897, *Tweedy 516* (HT: YU!; IT: NY!).

A. pacifica NAF **34:** 222. 1916. **CALIFORNIA. Siskiyou Co.:** Grenada, 19 Jun 1905, *Heller 8067* (HT: NY!; IT: DS!, US!).

A. palmeri NAF **34:** 221. 1916. **MEXICO. COAHUILA:** Saltillo, 10–20 Nov 1902, *Palmer 326* (HT: NY!; IT: US!).

A. puberula NAF **34:** 223. 1916. **CALIFORNIA:** Suisun Marsh, 6 Jul 1913, *Eastwood 3460* (HT: NY!; IT: CAS!, US!).

Agoseris altissima Mem. New York Bot. Gard. **1:** 458. 1900. **MONTANA:** Jack Creek, 14 Jul 1897, *Rydberg & Bessey 5289* (HT: NY!).

A. arachnoidea BTBC **28:** 511. 1901. **COLORADO:** Grayback mining camps and Placer Gulch, 25–27 Jun 1900, *Rydberg & Vreeland 5553* (HT: NY! on 2 sheets; IT: NEB, RM!).

A. attenuata BTBC **32:** 136. 1905. **COLORADO:** Mountains west of North Park, 27 Aug 1900, *Osterhout 2248* (HT: NY!; IT: NY!, RM! 3 sheets).

A. carnea Mem. New York Bot. Gard. **1:** 458. 1900. **Canada. BRITISH COLUMBIA:** Mt. Queest, 25 Jul 1889, *Macoun s.n.* (HT: NY!; IT: CAN).

A. humilis BTBC **32:** 137. 1905. **COLORADO:** Ironton Park, 9 miles south of Ouray, 11 Sep 1901, *Underwood & Selby 308* (HT: NY!).

A. leontodon var. *aspera* Mem. New York Bot. Gard. **1:** 457. 1900. **IDAHO:** Henry's Lake and Mt. Chauvet, 29 Jul 1897, *Rydberg & Bessey 5286* (HT: NY!; IT: US!).

A. leontodon var. *pygmaea* Mem. New York Bot. Gard. **1:** 458. 1900. **MONTANA:** Old Hollowtop near Pony, 9 Jul 1897, *Rydberg & Bessey 5288* (HT: NY!).

A. maculata BTBC **32:** 136. 1905. **COLORADO:** Silver Plume, 24 Aug 1895, *Shear 4605* (HT: NY!).

A. pubescens BTBC **28:** 511. 1901. **WYOMING. Sheridan Co.:** Big Horn Mts., Aug 1899, *Tweedy 2055* (HT, YU!).

A. roseata BTBC **32:** 136. 1905. **COLORADO:** Bear River, 20 miles below Steamboat Springs, 30 Aug 1899, *Osterhout s.n.* (HT, NY!).

A. rostrata BTBC **32:** 137. 1905. **COLORADO. Boulder Co.:** Lower Boulder Canyon, 25 Jun 1901, *Osterhout 2478* (HT: NY!; IT: RM! 2 sheets).

A. turbinata BTBC **38:** 20. 1911. **CANADA. ALBERTA:** Gravel slope of Tunnel Mt., 3 Jul 1899, *McCalla 2063* (HT: NY!).

A. villosa Mem. New York Bot. Gard. **1:** 458. 1900. **MONTANA:** Helena, Jul 1891, *Kelsey s.n.* (HT: NY!; IT: MONT dated 5 Jul and labeled Basin, MU).

Amauriopsis NAF **34:** 37. 1914. T: *Amauria dissecta* A. Gray.

Ambrosia californica NAF **33:** 20. 1922. **CALIFORNIA:** Santa Barbara, 1865, *Torrey 226* (HT: NY!; IT: NY!).

A. media BTBC **37:** 127. 1910. **COLORADO:** Fort Collins, 27 Aug 1885, *Crandall s.n.* (HT: NY!).

A. rugelii NAF **33:** 19. 1922. **FLORIDA:** 1845, *Rugel 508* (HT: NY!; IT: US!).

A. striata Brittonia **1:** 96. 1931. **COLORADO. Weld Co.:** Windsor, Aug 1901, *Osterhout 2338* (HT: NY!; IT: RM!).

A. variabilis Brittonia **1:** 97. 1931. **SOUTH DAKOTA:** Deadwood, 4 Aug 1913, *Carr 124* (HT: NY!).

Anaphalis angustifolia BTBC **37:** 326. 1910. **MONTANA:** Mt. MacDonald, Jul 1900, *Elrod s.n.* (HT: NY!).

Antennaria acuta BTBC **37:** 323. 1910. **CANADA. ALBERTA:** Near Lake O'Hara, 8 Aug 1904, *Macoun* (herbarium *#65423*) *s.n.* (HT: NY!; IT: CAN, US!).

A. anaphaloides Mem. New York Bot. Gard. **1:** 409. 1900. **MONTANA. Gallatin Co.:** Spanish Basin, 28 Jun 1897, *Rydberg & Bessey 5170* (HT: NY!; IT: CAN, MONT, NEB 2 sheets, RM!, US!).

A. angustifolia BTBC **26:** 546. 1899. **CALIFORNIA:** Yosemite Valley, 1865, *Torrey s.n.* (ST: NY!, US! numbered *256*); Hat Creek, *Newberry s.n.* (ST: NY!).

A. bracteosa Mem. New York Bot. Gard. **1:** 413. 1900. **MONTANA:** Jack Creek Canyon, 15 Jul 1897, *Rydberg & Bessey 5144* (HT: NY!; IT: CAN, MONT, NEB).

A. campestris BTBC **24:** 304. 1897. **NEBRASKA:** Lincoln, Jun 1887, *Webber s.n.* (HT: NY!—designated by Rydberg in herb.; IT: NEB).

A. flavescens Mem. New York Bot. Gard. **1:** 411. 1900. **MONTANA:** Bridger Mts., 18 [11 in publication] Jun 1897, *Rydberg & Bessey 5145* (HT: NY!; IT: CAN, MONT, NEB 2 sheets, RM!, US!).

A. foliacea var. *humilis* Mem. New York Bot. Gard. **1:** 414. 1900. **MONTANA:** Bridger Mts., 14 Jun 1897, *Rydberg & Bessey 5149* (HT: NY!; IT: MONT, NEB, RM!, US!).

A. microphylla BTBC **24:** 303. 1897. **MONTANA:** Manhattan, 17 Jul 1895, *Rydberg 2831* (HT: NY!; IT: NY!).

A. oblanceolata Mem. New York Bot. Gard. **1:** 409. 1900. **MONTANA:** Bridger Mts., 18 Jun 1897, *Rydberg & Bessey 5168* (HT: NY!; IT: CAN, MONT, NEB 2 sheets, RM!, US!).

A. piperi BTBC **28:** 21. 1901. **OREGON:** Mts. of northern Oregon, *Wilkes Expedition s.n.* (LT: NY!—Bayer, Taxon 37: 296. 1988).

A. rosulata BTBC **24:** 300. 1897. **ARIZONA:** Allen's Park, Mogollon Mts., 25 May 1887, *Mears 40* (HT: NY!—designated by Rydberg in herb.); IT: NY! not numbered).

A. sierrae-blancae BTBC **32:** 127. 1905. **COLORADO:** Sierra Blanca, 1877, *Hooker & Gray s.n.* (HT: GH!).

A. umbrinella BTBC **24:** 302. 1897. **MONTANA:** Little Belt Mts., Long Baldy, 19 Aug 1896, *Flodman 859* (HT: NY!; IT: NY!, US).

Antheropeas NAF **34:** 97. 1915. T: *Bahia wallacei* A. Gray.

A. australe NAF **34:** 98. 1915. **MEXICO.** Mountains of northern Lower California, 7 Jul 1884, *Orcutt s.n.* (HT: NY!; IT: NY!).

A. tenuifolium NAF **34:** 99. 1915. **MEXICO. LOWER CALIFORNIA:** San Sebastian, 28 Apr 1889, *Brandegee s.n.* (HT: US!; CT: NY! 2 sheets).

Arnica alata NAF **34:** 342. 1927. **CALIFORNIA:** Yosemite Valley and Mts., 1865, *Torrey 258a* (HT: NY!; IT: NY!).

A. amplifolia NAF **34:** 354. 1927. **IDAHO:** Bear Creek, Bitter Root Forest Reserve, 26 Aug 1897, *Leiberg 2934* (HT: NY!).

A. arachnoidea NAF **34:** 353. 1927. **UTAH:** Big Cottonwood Canyon, below Silver Lake, 4 Jul 1915, *Rydberg & Carlton 6609* (HT: NY!; IT: CAN).

A. austinae NAF **34:** 340. 1927. **CALIFORNIA:** Lake City Canyon, Jul 1898, *Austin & Bruce 2165* (HT: NY!; IT: NY! 2 sheets, UC).

A. borealis NAF **34:** 351. 1927. **ALASKA:** Juneau, 10 Jul 1917, *Anderson 614* (HT: NY!).

A. brevifolia NAF **34:** 329. 1927. **ALASKA:** Chitaslene Glacier, Copper River region, 19 Jun 1902, *Potg 46* (HT: US).

A. bruceae NAF **34:** 347. 1927. **CALIFORNIA. Modoc Co.:** Lake Annie, Jun 1898, *Bruce 2174* in part (HT: NY!; IT: DS!).

A. caudata BTBC **37:** 463. 1910. **UTAH. Salt Lake Co.:** Big Cottonwood Canyon, near Lake Catherine, 3 Aug 1905, *Garrett 1547* (HT: NY!; IT: RM!, UT!).

A. chandleri NAF **34:** 339. 1927. **CALIFORNIA:** Hupi Indian Reservation, Jun 1901, *Chandler 1298* (HT: NY!; IT: UC, US!).

A. ciliaris NAF **34:** 351. 1927. **CALIFORNIA. Placer Co.:** Railroad Canyon on road to Tahoe, 2 Aug 1896, *Sonne s.n.* (HT: NY!).

A. coloradensis BTBC **32:** 131. 1905. **COLORADO:** 1872, *Parry s.n.* (HT: NY!; IT: NEB, NY! numbered *902*).

A. cusickii NAF **34:** 343. 1927. **OREGON:** Cascade Mts., 11 Jul 1902, *Cusick 2873* (HT: NY!; IT: MO, ORE, POM, UC, US!).

A. eastwoodiae NAF **34:** 343. 1927. **CALIFORNIA. Del Norte Co.:** Gasquet, French Hill, 14 Sep 1912, *Eastwood 2211* (HT: NY!; IT: NY!, US!).

A. elongata NAF **34:** 350. 1927. **CANADA. BRITISH COLUMBIA:** Chilliwack River, 25 Jul 1906, *Spreadborough* (herbarium *#77007*) *s.n.* (HT: CAN—fide Shchepanek in litt. 1984; IT: NY!).

A. fernaldii NAF **34:** 333. 1927. **CANADA. NEWFOUNDLAND:** Table Mt., 16–17 Jul 1914, *Fernald & St. John 10875* (HT: NY!; IT: CAN, GH, RM!).

A. filipes E. L. Greene ex Rydb., NAF **34:** 350. 1927. **CANADA. BRITISH COLUMBIA:** Selese Creek, Chilliwack River, 4 Aug 1906, *Spreadborough* (herbarium *#77025*) *s.n.* (HT: NY!; IT: CAN).

A. flodmanii NAF **34:** 334. 1927. **MONTANA:** Madison Range, Spanish Peaks, 14 Jul 1896, *Flodman 898* (HT: NY! on 2 sheets; IT: MO, US!).

A. glabrata NAF **34:** 335. 1927. **OREGON:** Crater Lake, 10 Aug 1898, *Austin & Bruce 1627* (HT: NY!; IT: US! dated 10 Aug 1897).

A. gracilis BTBC **24:** 297. 1897. **MONTANA:** Madison Range, Spanish Peaks, 14 Jul 1896, *Flodman 901* (HT: NY! on 3 sheets; IT: US!).

A. granulifera Fl. Rocky Mts. 978. 1917 [1918]. **MONTANA:** Little Belt Mts., Long Baldy, 19 Aug 1896, *Flodman s.n.* (LT: NY!—Rydberg, NAF **34:** 335. 1927).

A. hirticaulis NAF **34:** 349. 1927. **WASHINGTON. Skamania Co.:** Crooks, 7 Sep 1909, *Suksdorf s.n.* (HT, NY!).

A. humilis NAF **34:** 341. 1927. **CANADA. ALBERTA:** Lake Louise, 20 Jul 1904, *Macoun* (herbarium *#65504*) *s.n.* (HT: NY!; IT: CAN).

A. illiamnae NAF **34:** 331. 1927. **ALASKA:** Lake Illiamna region, 20 Jul 1902, *Gorman 163* (HT: US; CT: NY!).

A. jonesii Fl. Rocky Mts. 979. 1917 [1918]. **UTAH:** Alta, 31 Jul 1879, *Jones 1119* (LT: NY!—Rydberg, NAF **34:** 335. 1927; ILT: NY!, POM, UTC!).

A. kodiakensis NAF **34:** 353. 1927. **ALASKA:** Kodiak Island, 1910, *Mylroie 30* (HT: NY!).

A. leptocaulis NAF **34:** 336. 1927. **CANADA. [BRITISH COLUMBIA:]** Vancouver Island, Mt. Mark, 25 Jul 1887, *Macoun s.n.* (HT: NY!; IT: CAN, NY! numbered *14750*, US!).

A. membranacea NAF **34:** 338. 1927. **OREGON. Jackson Co.:** Wimmer, 13 Jun 1892, *Hammond 231* (HT: NY!; IT: US! dated 4 May 1889, WS).

A. mendenhallii NAF **34:** 329. 1927. **CANADA.** Old Man's Creek, a branch of the Koyukuk River, 8 Jul 1901, *Mendenhall s.n.* (HT: US).

A. monocephala Mem. New York Bot. Gard. **1:** 435. 1900. **MONTANA:** Bridger Mts., 14 Jun 1897, *Rydberg & Bessey 5221* (HT: NY!; IT: CAN, NEB, US!).

A. nutans NAF **34:** 328. 1927. **ALASKA:** Near Port Clarence, 30 Jul 1901, *Walpole 1618* (HT: NY!; IT: US!).

A. oligolepis NAF **34:** 336. 1927. **CANADA. BRITISH COLUMBIA:** Hazelton, Skeena River, 23 Jun 1917, *Macoun* (herbarium *#96048*) *s.n.* (HT: NY!; IT: CAN).

A. ovalis NAF **34:** 338. 1927. **CANADA.** Rocky Mountains, Crow Nest Pass, 31 Jul 1897, *Macoun* (herbarium *72719*) *s.n.* (HT: CAN—fide Shchepanek in litt. 1984).

A. paucibracteata NAF **34:** 336. 1927. **WYOMING. Albany Co.:** Medicine Bow Mts., 3 Aug 1900, *Nelson 7941* (HT: NY! on 2 sheets; IT: MO, POM, RM! US!).

A. pedunculata BTBC **24:** 297. 1897. **MONTANA:** Spanish Basin, 11 Jul 1896, *Flodman 899* (HT: NY!—designated by Rydberg in herb.).

A. puberula Fl. Rocky Mts. 979. 1917 [1918]. **CANADA. ALBERTA:** Lake Louise, 22 Jul 1904, *Macoun* (herbarium *#65523*) *s.n.* (LT: NY!—Rydberg, NAF **34:** 333. 1927; ILT: CAN, US!).

A. pumila Mem. New York Bot. Gard. **1:** 433. 1900. **COLORADO:** Gray's Peak, Aug & Sep 1872, *Torrey s.n.* (HT: NY!).

A. sancti-laurentii NAF **34:** 328. 1927. **[U.S.S.R.]. SIBERIA:** St. Laurence Bay, *Chamisso s.n.* (HT: GH!).

A. sanhedrensis NAF **34:** 342. 1927. **CALIFORNIA. Lake Co.:** Mt. Sanhedrin, between Potter Valley & Hulville, 25 Jul 1902, *Heller 5985* (HT: NY!; IT: POM, RM!, UC, US!).

A. sulcata NAF **34:** 344. 1927. **CALIFORNIA. Siskiyou**

Co.: Scott Mts., 22 Aug 1876, *Greene 1005* (HT: GH!).

A. tenuis BTBC **28:** 20. 1901. **Wyoming. Sheridan Co.:** Big Horn Mts., Aug 1899, *Tweedy 2094* (HT: NY!; IT: YU!).

A. tomentulosa BTBC **28:** 20. 1901. **Wyoming:** Buffalo Fork, Aug 1897, *Tweedy 523* (HT: NY!; IT: YU! dated 15 Jul–15 Aug).

A. wilsonii NAF **34:** 332. 1927. **CANADA. Ontario:** West of James Bay, 140 miles up Kapiscoro River, 15 Jul 1902, *Wilson* (herbarium *#54043*) *s.n.* (HT: CAN—fide Shchepanek in litt. 1984).

Artemisia alaskana NAF **34:** 281. 1916. **Alaska:** Yukon River between Nulato and Nowikakat, 23–27 Jul 1889, *Russell s.n.* (HT: US!).

A. argophylla NAF **34:** 274. 1916. **Colorado:** Longs Peak, 14 Aug 1907, *Clements s.n.* (HT: NY!).

A. bourgeauana BTBC **37:** 454. 1910. **CANADA. Saskatchewan:** 1857–1859, *Bourgeau s.n.* (HT: NY!).

A. brittonii BTBC **32:** 129. 1905. **Colorado:** Golden, 8 Oct 1882, *Britton s.n.* (HT: NY!).

A. candicans BTBC **24:** 296. 1897. **Montana:** Little Belt Mts., 18 Aug 1896, *Flodman 882* (HT: NY! on 2 sheets, one labeled 6 miles from Barker, the other 9 miles; IT: US! labeled 6 miles from Barker).

A. comata NAF **34:** 263. 1916. **Alaska:** Camden Bay, Collinson Point, 17 Jul 1914, *Johansen* (herbarium *#89994*) *122* (HT: CAN—fide Shchepanek in litt. 1985; IT: CAN, NY!).

A. cooleyae NAF **34:** 265. 1916. **Alaska:** Juneau, Silver Bow Basin, 6 Aug 1891, *Cooley s.n.* (HT: CAN—fide Shchepanek in litt. 1984; IT: US!).

A. cuneata NAF **34:** 269. 1916. **Colorado:** Idaho Springs, 27 Aug 1895, *Shear 4617* (HT: NY!).

A. diversifolia BTBC **28:** 20. 1901. **Idaho:** Forest between Priest Lake & East Fork, 27 Jul 1900, *MacDougal 190* (HT: NY!).

A. dracunculoides var. *wolfii* BTBC **32:** 128. 1905. **Colorado:** Twin Lakes, 1873, *Wolf 530* (HT: NY!; IT: NY!).

A. falcata NAF **34:** 271. 1916. **South Dakota:** Fort Pierre, 29 Sep 1897, *Griffiths 757* (HT: US!).

A. floccosa BTBC **24:** 297. 1897. **Montana:** Lima, 6 Aug 1895, *Rydberg 2942* (HT: NY!; IT: NY!, US!).

A. flodmanii NAF **34:** 276. 1916. **Montana:** Little Belt Mts., 9 miles east of Barker, 18 Aug 1896, *Flodman 883* (HT: NY!; IT! NY!).

A. ghiesbreghtii NAF **34:** 271. 1916. **MEXICO. Chiapas:** *Ghiesbreght 1515* [or perhaps *155, 1551* in publication] (HT: NY!).

A. gormanii NAF **34:** 267. 1916. **Alaska:** Lake Iliamna south of Iliama Point, 1 Sep 1902, *Gorman 281* (HT: US!).

A. gracillima NAF **34:** 253. 1916. **New Mexico. Sierra Co.:** 1 mile west of Hillsboro, 2 Oct 1904, *Metcalfe 1444* (HT: NY!; IT: CAS!, US!).

A. graveolens BTBC **24:** 296. 1897. **Montana:** Little Belt Mts., Yogo Baldy, 24 Aug 1896, *Flodman 881* (HT: NY!; IT: NY!, US! both as Long Baldy and dated 19 Aug. Long Baldy was cited in the original publication but the Yogo Baldy sheet has TYPE in Rydberg's hand).

A. herriotii BTBC **37:** 455. 1910. **CANADA. Alberta:** Edmonton, 25 Aug 1906, *Macoun & Herriot* (herbarium *#72825*) *s.n.* (HT: NY!; IT: CAN).

A. hyperborea NAF **34:** 262. 1916. **CANADA. Northwest Territory:** Bernard Harbor, 14 Aug 1915, *Johansen 304a* [304 in publication] (HT: CAN—fide Shchepanek in litt. 1984; IT: NY! 2 sheets).

A. leibergii NAF **34:** 267. 1916. **WASHINGTON. Franklin Co.:** Fish Hook Ferry, 18 Sep 1894, *Leiberg 935* (HT: NY!; IT: US!).

A. maccallae NAF **34:** 254. 1916. **[CANADA]. Alberta:** Banff, Tunnel Mt., 16 Sep 1899, *McCalla 2016* (HT: NY!; IT: US!).

A. manca NAF **34:** 256. 1916. **Alaska:** Cold Bay, 31 Jul 1904, *Piper 4223* (HT: US!).

A. minuta NAF **34:** 261. 1916. **CANADA. Alberta:** 50 miles north of Medicine Hat, May 1914, *Seton s.n.* (HT: NY!).

A. neomexicana E. L. Greene ex Rydb., NAF **34:** 279. 1916. **New Mexico. Grant Co.:** Hillsboro Peak, Black Range, 25 Aug 1904, *Metcalfe 1248* (HT: NY!; IT: RM!, UTC!).

A. obtusa NAF **34:** 274. 1916. On the Columbia, *Scouler 234* (HT: NY!).

A. platyphylla NAF **34:** 275. 1916. **Washington. Spokane Co.:** Spokane, Spokane River, Jun 1897, *Elmer 867* (HT: NY!; IT: US!).

A. pudica BTBC **32:** 130. 1905. **Colorado:** Gunnison, 24 Jul 1901, *Baker 573* (HT: NY!; IT: RM!, US!).

A. revoluta NAF **34:** 272. 1916. **MEXICO. Chihuahua:** Chihuahua, 4 Nov 1885, *Pringle 290* (HT: NY!; IT: NY!).

A. ripicola NAF **34:** 256. 1916. **Oregon. Sherman Co.:** Biggs, 31 May 1910, *Heller 10117* (HT: NY!; IT: CAS!).

A. sulcata NAF **34:** 270. 1916. **MEXICO. Chihuahua:** Casas Grandes, 30 Aug 1899, *Townsend & Barber 437* (HT: NY!; IT: US!).

A. tacomensis NAF **34:** 262. 1916. **Washington:** Mt. Rainier, 27 Aug 1896, *Flett s.n.* (HT: US!).

A. tenuis Mem. New York Bot. Gard. **1:** 431. 1900. **Montana:** Emigrant Gulch, 23 Aug 1897, *Rydberg & Bessey 5201* (HT: NY!; IT: NEB, US!).

A. tenuis var. *integerrima* Mem. New York Bot. Gard. **1:** 432. 1900. **Montana:** Emigrant Gulch, 23 Aug 1897, *Rydberg & Bessey 5201a* (HT: NY!; IT: US!).

A. texana NAF **34:** 274. 1916. **Texas:** Colorado, 16 Oct 1902, *Tracy 8141* (HT: US!; IT: NY!).

A. tyrrellii NAF **34:** 262. 1916. **CANADA. British Columbia:** Selkirk Trail, 6 Sep 1898, *Tyrrell* (herbarium *#19439*) *s.n.* (HT: CAN—fide Shchepanek in litt. 1984; CT: NY!).

A. unalaskensis NAF **34:** 266. 1916. **[Alaska]. Unalaska:** Bering Sea, 22 Aug 1891, *Macoun* (herbarium *#20625*) *s.n.* (HT: CAN—fide Shchepanek in litt. 1984).

A. underwoodii BTBC **32:** 129. 1905. **Colorado:** Southeast of Ouray, 7 Sep 1901, *Underwood & Selby* 74 (HT: NY!).

A. vaseyana NAF **34:** 283. 1916. **Washington:** 1889, *Vasey 480* (HT: NY!; IT: US!).

Artemisiastrum NAF **34:** 285. 1916. T: *Artemisia palmeri* A. Gray.

Aster amplexifolius Mem. New York Bot. Gard. **1:** 391.

1900. **Montana:** Headwaters of Jocko River, Jul 1883, *Canby 170* (HT: NY!).

A. batesii Brittonia **1:** 102. 1931. **Nebraska:** St. Paul, 25 Sep 1910, *Bates 5440* (HT: NY!; IT: NEB not numbered).

A. brevibracteatus Mem. New York Bot. Gard. **1:** 392. 1900. **Washington. Spokane Co.:** Spangle, 10 Aug 1889, *Suksdorf 928* (HT: NY!; IT: US!).

A. butleri BTBC **37:** 139. 1910. **Montana:** Gateway, 16 Aug 1908, *Butler 443* (HT: NY!).

A. ciliomarginatus Mem. New York Bot. Gard. **1:** 392. 1900. **Montana:** Electric Peak, 20 Aug 1897, *Rydberg & Bessey 5128* (HT: NY!; IT: NEB, US!).

A. corymbiformis BTBC **31:** 655. 1904 [1905]. **Colorado:** West Cliff, 13 Aug 1896, *Shear 3460* (HT, NY!).

A. crassulus BTBC **28:** 504. 1901. **Colorado:** La Veta, 26 Sep 1900, *Vreeland 690a* (HT: NY!; IT: RM! numbered *690*).

A. eriocaulis BTBC **37:** 143. 1910. **Idaho. Kootenai Co.:** Valley of Traille River, 9 Aug 1892, *Sandberg, MacDougal & Heller 877* (HT: NY!; IT: CAS!, DS!).

A. finkii Brittonia **1:** 102. 1931. **Iowa:** Fayette, Sep 1895, *Fink s.n.* (HT: ISC—fide Qualls in litt. 1984).

A. griseolus BTBC **31:** 653. 1904 [1905]. **Colorado:** Mt. Harvard, 1896, *Clements 37* (HT: NY!; IT: NEB).

A. junciformis BTBC **37:** 142. 1910. **Montana:** East Gallatin Swamps, 24 Jul 1896, *Flodman 833* (HT: NY!; IT: US!).

A. laeviformis Brittonia **1:** 103. 1931. **Minnesota. Lincoln Co.:** Lake Benton, Aug 1891, *Sheldon S1335* (HT: MIN—fide Ownbey in litt. 1984; IT: NY! not numbered).

A. maccallae BTBC **37:** 138. 1910. **CANADA. Alberta:** Sulphur Mt., vicinity of Banff, 16 Aug 1899, *McCalla 2026* (HT: NY!).

A. macounii BTBC **37:** 136. 1910. **CANADA. Northwest Territory:** West of Old Wives Lake, Jul 1880, *Macoun s.n.* (HT: NY!).

A. mearnsii BTBC **37:** 144. 1910. **Wyoming:** Mammoth Hot Springs, 11 Oct 1902, *Mearns 4806* (HT: NY!; IT: US! no data).

A. mollis BTBC **28:** 22. 1901. **Wyoming. Sheridan Co.:** Big Horn Mts., 20 Aug 1899, *Tweedy 2029* (HT: NY!; IT: WS, YU!).

A. osterhoutii BTBC **31:** 654. 1904 [1905]. **Colorado. Weld Co.:** New Windsor, 29 Sep 1899, *Osterhout 2506* (HT: NY!; IT: NY! dated 9 Sep, RM! 2 sheets dated 25 Sep).

A. oticus Brittonia **1:** 103. 1931. **Minnesota:** Red Wing, Aug 1885, *Sandberg 3758* (HT: MIN—fide Ownbey in litt. 1984).

A. phyllodes BTBC **37:** 145. 1910. **Nebraska. Kimball Co.:** Lodge Pole Creek near Kimball, 12 Aug 1891, *Rydberg 165* (HT: NY!).

A. roseolus BTBC **37:** 141. 1910. **Montana:** Melrose, 1 Aug 1895, *Rydberg 2817* (HT: NY!).

A. scribneri Mem. New York Bot. Gard. **1:** 393. 1900. **Montana:** Bird Tail Creek, 6 Sep 1883, *Scribner 91* (HT: NY!).

A. subcaudatus BTBC **37:** 143. 1910. **CANADA. Alberta:** Laggan, Lake Louise region, 23 Aug 1902, *Rosendahl 1089* (HT: NY!).

A. subracemosus Mem. New York Bot. Gard. **1:** 394. 1900. **Montana:** Helena, 8 Sep 1890, *Kelsey s.n.* (HT: NY!).

A. subsalignus BTBC **37:** 140. 1910. **Colorado. Garfield Co.:** Glenwood Springs, 18 Aug 1906, *Osterhout 3397* (HT: NY!; IT: RM!).

A. tweedyi BTBC **31:** 655. 1904 [1905]. **Wyoming:** Copperton, 6 Aug 1901, *Tweedy 4096* (HT: NY!; IT: US!, YU!).

A. umbachii BTBC **37:** 144. 1910. **Montana:** Lake McDonald, 20 Aug 1901, *Umbach 275* (HT: NY!).

A. underwoodii BTBC **31:** 654. 1904 [1905]. **Colorado:** Ironton Park, 9 miles south of Ouray, 11 Sep 1901, *Underwood & Selby 318* (HT: NY!).

A. williamsii BTBC **37:** 136. 1910. **Wyoming:** Big Horn Mts., north fork of Clear Creek, 12 Aug 1898, *Williams s.n.* (HT: NY!).

A. wilsonii BTBC **37:** 138. 1910. **CANADA. British Columbia:** Armstrong, 1904, *Wilson 419* (HT: NY!).

A. woldenii Brittonia **1:** 103. 1931. **Iowa:** Estherville, 6 Sep 1924, *Wolden 1068* (HT: NY!; IT: ISC).

Asterigeron Fl. Rocky Mts. 891, 1067. 1917 [1918]. T: *Aster watsonii* A. Gray.

Bahia aristata NAF **34:** 36. 1914. **MEXICO. San Luis Potosí:** Hacienda de Angostura, 11 Jul 1891, *Pringle 5127* (HT: GH!).

B. ehrenbergii Schultz-Bip. ex Rydb., NAF **34:** 35. 1914. **MEXICO.** *Ehrenberg 362* (HT: GH!).

Baileya australis NAF **34:** 11. 1914. **MEXICO. Durango:** Santiago Papasquiaro, Apr & Aug 1896, *Palmer 50* (HT: NY!).

B. thurberi NAF **34:** 10. 1914. **Texas:** Along the Pecos, 1850, *Thurber 132* (HT: NY!).

Balsamorrhiza floccosa BTBC **27:** 629. 1900. **Montana. Gallatin Co.:** Spanish Basin, 1 Jul 1897, *Rydberg & Bessey 5175* (HT: NY!; IT: MONT, NEB 2 sheets, US!).

B. tomentosa BTBC **27:** 628. 1900. **Wyoming:** Big Horn Mts., headwaters of the Tongue River, Jul 1898, *Tweedy 10* (HT: NY!; IT: YU!).

Bidens filamentosa Brittonia **1:** 104. 1931. **Minnesota:** Montevideo, 11 Sep 1904, *Moyer & Nisbit s.n.* (HT: NY!).

Boebera ciliosa NAF **34:** 167. 1915. **MEXICO. Durango:** Durango, Apr–Nov 1896, *Palmer 920* (HT: NY!; IT: US!).

Carduus butleri BTBC **37:** 542. 1910. **Montana:** Big Fork, 28 Jul 1908, *Butler 674* (HT: NY!).

C. canovirens Mem. New York Bot. Gard. **1:** 450. 1900. **Montana:** Jack Creek Canyon, 15 Jul 1897, *Rydberg & Bessey 5213* (HT: NY!; IT: US!).

C. centaureae BTBC 28: 507. 1901. **Colorado. Laramie [Larimer] Co.:** 19 Aug 1895, *Cowan s.n.* (HT: NY! on 2 sheets).

C. coloradensis BTBC **32:** 132. 1905. **Colorado:** Pagosa Springs, 28 Jul 1899, *Baker 644* (HT: NY!; IT: US!).

C. erosus BTBC **28:** 507. 1901. **Colorado:** Durango, 25 Jul 1896, *Tweedy 517* (HT: US!; IT: YU! as La Plata Mts.; CT: NY!).

C. floccosus BTBC **32:** 133. 1905. **Colorado. Eagle Co.:** Wolcott, 11 Jul 1902, *Osterhout 2652* (HT: NY!; IT: RM! 2 sheets).
C. flodmanii Mem. New York Bot. Gard. **1:** 451. 1900. **Montana:** East Gallatin Swamps, 24 Jul 1896, *Flodman 879* (HT: NY! on 2 sheets; IT: US!).
C. griseus BTBC **28:** 509. 1901. **Colorado:** Telluride, Aug 1894, *Tweedy 321* (HT: US!; IT: NY!, YU! numbered 231).
C. kelseyi Mem. New York Bot. Gard. **1:** 449. 1900. **Montana:** Main range of Rocky Mts., 1891 and 1892, *Kelsey s.n.* (HT: NY! dated July 1891; IT?: MU dated 21 Jul 1892 and numbered *227*).
C. lacerus BTBC **37:** 543. 1910. **Utah. Wasatch Co.:** Near Midway, 6 Jul 1905, *Carlton & Garrett* 6732 (HT: NY!; IT: RM!).
C. oblanceolatus BTBC **28:** 510. 1901. **Colorado:** Twin Lakes, 1896, *Clements 385* (HT: NY!; IT: NEB!).
C. olivescens BTBC **37:** 544. 1910. **Utah:** Aquarius Plateau, 5 Aug 1905, *Rydberg & Carleton 7450* (HT: NY!).
C. oreophilus BTBC **28:** 509. 1901. **Colorado:** Georgetown, Aug 1895, *Rydberg s.n.* (HT: NY!).
C. osterhoutii BTBC **32:** 131. 1905. **Colorado. Eagle Co.:** Red Cliff, 17 Jul 1902, *Osterhout 2706* (HT: NY!).
C. perplexans BTBC **32:** 132. 1905. **Colorado:** Cimarron, 29 Jun 1901, *Baker 286* (HT: NY!; IT: RM!).
C. plattensis Contr. U.S. Natl. Herb. **3:** 167. 1895. **Nebraska. Thomas Co.:** Plummer Ford, Dismal River, 5 Jul 1893, *Rydberg 1356* (HT: US!; IT: NEB, NY!).
C. plattensis var. *spinosior* Contr. U.S. Natl. Herb. **3:** 167. 1895. **Nebraska. Box Butte Co.:** 7 Jul 1892, *Smith & Pound 64* (ST: NEB—fide Bolick in litt. 1984, NY!); **Holt Co.:** Paddock, 25 Jul 1893, *Clements 2963* (*2964* in publication) (ST: NEB—fide Bolick in litt. 1984, US!).
C. polyphyllus BTBC **37:** 542. 1910. **Montana:** Mountains near Indian Creek, 21 Jul 1897, *Rydberg & Bessey 5216* (HT: NY!; IT: US!).
C. pulchellus E. L. Greene ex Rydb., Fl. Colorado 400, 401. 1906. **Colorado:** Piedra, 13 Jul 1899, *Baker 646* (HT: NY!; IT: US!).
C. pulcherrimus BTBC **28:** 510. 1901. **Wyoming:** Big Horn Mts., headwaters of Clear Creek and Crazy Woman River, 20 Jul–15 Aug 1900, *Tweedy 3048* (HT: NY!; IT: RM!, US!, YU!).
C. tracyi BTBC **32:** 133. 1905. **Colorado:** Mancos, 24 Jun 1898, *Baker, Earle & Tracy 90* (HT: NY!; IT: NEB, RM! 2 sheets, US!).
C. truncatus E. L. Greene ex Rydb., Fl. Colorado 400, 401. 1906. **Colorado:** West of Mancos, 8 Jul 1898, *Baker, Earle & Tracy 401* (HT: NY!; IT: NEB, NY!; RM! 2 sheets, US!).
C. tweedyi Mem. New York Bot. Gard. **1:** 449. 1900. **Montana. Park Co.:** Aug 1887, *Tweedy 350* (HT: YU!; IT: NY!).
Chaenactis aurea E. L. Greene ex Rydb., NAF **34:** 68. 1914. **California. Los Angeles Co.:** Wilson's Peak, Jul 1901, *Abrams 1890* (HT: NDG!; IT: DS!)
C. humilis NAF **34:** 72. 1914. **Idaho:** Bear River Range, Franklin Basin, 24 Jul 1910, *C. P. Smith 2273* (HT: NY!).
C. rubricaulis NAF **34:** 72. 1914. **California. Placer Co.:** Deer Park, 22 Jun 1910, *Walker 2170* (HT: NY!; IT: NY!, RM!, UC, US!).
Chamaechaenactis BTBC **33:** 155. 1906. T: *Chaenactis scaposa* Eastw.
Chamartemisia NAF **34:** 242. 1916. T: *Tanacetum compactum* H. M. Hall
Chrysopsis alpicola BTBC **31:** 649. 1904 [1905]. **Colorado:** Clark's Peak, 1 Aug 1896, *Baker s.n.* (HT: NY!; IT: CAS!, DS! 2 sheets, NY!, UC).
C. amplifolia BTBC **31:** 648. 1904 [1905]. **Colorado. Boulder Co.:** Longmont, Jul 1902, *Tweedy 4898* (HT: NY!).
C. angustifolia BTBC **37:** 128. 1910. **Nebraska. Hooker Co.:** Middle Loup River, near Mullen, 14 Sep 1893, *Rydberg 1766* (HT: NY! on 2 sheets; IT: US).
C. arida A. Nels. ex Rydb., Fl. Colorado 339, 340. 1906. **Wyoming:** Tie City, 20 Jul 1900, *Nelson 8224a* (HT: NY!).
C. ballardii Brittonia **1:** 100. 1931. **Minnesota:** Chaska, Jul 1891, *Ballard 640* (HT: MIN—fide Ownbey in litt. 1984).
C. barbata BTBC **37:** 130. 1910. **Idaho. Kootenai Co.:** Valley of Spokane River, 17 Jul 1893, *Sandberg, MacDougal & Heller 664* (HT: NY!; IT: US).
C. butleri BTBC **37:** 129. 1910. **Montana:** Gateway, 17 Aug 1908, *Butler 620* (HT: NY!).
C. caudata BTBC **31:** 648. 1904 [1905]. **Colorado:** Ruxton Dell, 2 Aug 1901, *Clements & Clements 143* (HT: NY!; IT: NEB, US).
C. depressa Mem. New York Bot. Gard. **1:** 381. 1900. **Wyoming:** Yellowstone Park, lower Geyser Basin, 4 Aug 1897, *Rydberg & Bessey 5067* (HT: NY!; IT: CAN, NY!, US).
C. grandis BTBC **37:** 129. 1910. **Montana:** Jocko Creek, 16 Jun 1901, *MacDougal 275* (HT: NY!).
C. horrida BTBC **31:** 648. 1904 [1905]. **Colorado. Weld Co.:** New Windsor, 8 Aug 1900, *Osterhout 2326* (HT: NY!).
Chrysothamnus filifolius BTBC **28:** 503. 1901. **Colorado:** Granite, 1896, *Clements 390* (HT: NY!).
C. marianus BTBC **37:** 131. 1910. **Utah:** Along Sevier River, below Marysvale, 20 Jul 1905, *Rydberg & Carlton 6993* (HT: NY!; IT: RM!, US).
C. newberryi BTBC 31: 652. 1904 [1905]. **New Mexico:** Canyon Largo, 15 Sep 1859, *Newberry s.n.* (HT: NY!).
C. patens BTBC **31:** 652. 1904 [1905]. **Colorado:** 1873, *Wolf 452* (HT: NY!).
C. salicifolius BTBC **37:** 130. 1910. **Utah:** Strawberry Valley, Sep 1883, *Leonard 288* (HT: NY!; IT: BRY).
C. scoparius BTBC **28:** 504. 1901. **Colorado:** LaVeta, 9 Oct 1900, *Vreeland 698* (HT: NY!; IT: NY!; RM!).
C. stenolepis BTBC **37:** 131. 1910. **Montana:** Pass Creek, near Bridger Pass, *H. Engelmann s.n.* (HT: NY!).
Cirsium lactucinum Fl. Rocky Mts. 1010, 1068. 1917 [1918]. **Utah:** San Juan River near Bluffs, 29 Aug 1911, *Rydberg 10001* (HT: NY!; IT: US!, UT!).
C. subniveum Fl. Rocky Mts. 1006, 1068. 1917 [1918].

WYOMING: Jackson Hole, 21 Aug 1894, *Nelson 1070* (HT: US!; IT: RM! dated 22 Aug).

Clomenocoma laciniata NAF **34:** 166. 1915. ARIZONA: Hackberry, 24 May 1884, *Jones s.n.* (HT: NY!).

Coleosanthus albicaulis BTBC **31:** 646. 1904 [1905]. COLORADO: Manitou, 1900, *Clements s.n.* (HT, NY!).

Crepis angustata BTBC **32:** 135. 1905. COLORADO: North Park, 28 Jul 1896, *Baker s.n.* (HT: NY!; IT: NY!).

C. denticulata BTBC **32:** 135. 1905. COLORADO: Lake John, North Park, 19 Aug 1898, *Shear & Bessey 4004* (HT: NY!).

C. glaucella BTBC **28:** 512. 1901. WYOMING: Pacific Creek, Sep 1897, *Tweedy 603* (HT: NY!; IT: YU!).

C. perplexans BTBC **32:** 135. 1905. WYOMING. **Carbon Co.:** Encampment, 15 Jun 1901, *Tweedy 4081* (HT: NY!; IT: YU!).

C. petiolata BTBC **32:** 134. 1905. COLORADO: Along Bear River, 5 miles east of Hayden, 31 Aug 1899, *Osterhout 21* (HT: NY!; IT: RM! numbered *1785*).

C. pumila Mem. New York Bot. Gard. **1:** 462. 1900. MONTANA: Bridger Mts., 14 Jun 1897, *Rydberg & Bessey 5305* (HT: NY!; IT: US! 2 sheets).

C. runcinata var. *alpicola* BTBC **24:** 299. 1897. MONTANA: Little Belt Mts., Yogo Baldy, 24 Aug 1896, *Flodman 931* (HT: NY! on 3 sheets).

C. seselifolia BTBC **38:** 14. 1911. IDAHO. **Nez Perces Co.:** Valley of Big Potlatch River, 6 Jun 1892, *Sandberg, MacDougal & Heller 326* (HT: NY!).

C. tomentulosa BTBC **32:** 134. 1905. COLORADO: Ruxton Dell, 18 Jul 1901, *Clements & Clements 342* (HT: NY!; IT: DS!, GH, MO, NEB, RM!, US!).

Crossostephium insulare NAF **34:** 244. 1916. CALIFORNIA: San Clemente Island, Pot's Trail, Jun 1903, *Trask 286* (HT: NY!; IT: US!).

Cyclachaena lobata NAF **33:** 10. 1922. MEXICO. NUEVO LEÓN: Monterey, Aug 1911, *Arsène 6103* (HT: US!; CT: NY!).

C. pedicellata NAF **33:** 10. 1922. NEW MEXICO: Santa Fe, 14 Sep 1895, *Mulford 1284* (HT: NY!).

Dicoria hispidula NAF **33:** 12. 1922. CALIFORNIA: Whitewater Desert, 1881, *Parish & Parish 7* (HT: US!).

D. oblongifolia NAF **33:** 12. 1922. [CALIFORNIA.] Southern California, 1875, *Palmer 636* (HT: US!).

Dugaldia grandiflora NAF **34:** 120. 1915. CALIFORNIA: 1868 or 1869, *Kellogg & Harford 469* (HT: NY!; IT: GH, MO, US!).

Erigeron conspicuus Mem. New York Bot. Gard. **1:** 400. 1900. MONTANA: Jack Creek Canyon, 14 Jul 1897, *Rydberg & Bessey 5095* (LT: NY!—Cronquist, Brittonia **6:** 156. 1947; ILT: US!).

E. earlei BTBC **32:** 126. 1905. COLORADO: Antonito, 19 Jul 1898, *Earle s.n.* (HT: NY!).

E. evermannii Fl. Rocky Mts. 903, 1067. 1917 [1918]. IDAHO: Top of Mount Parks, 15 Aug 1895, *Evermann 378* (HT: US!).

E. flabellifolius BTBC **26:** 545. 1899. WYOMING: Yount's Peak, Teton Forest Reserve, Aug 1897, *Tweedy 536* (HT: NY!; IT: YU!).

E. fruticetorum Fl. Rocky Mts. 906, 1067. 1917 [1918]. UTAH: La Sal Mts., north of Brumley Creek, 14 Jul 1911, *Rydberg & Garrett 8912* (HT: NY!).

E. gracilis Mem. New York Bot. Gard. **1:** 404. 1900. WYOMING: Yellowstone Park, Slough Creek, Jul 1885, *Tweedy 702* (HT: YU!; IT: NY!, US!).

E. incanescens BTBC **28:** 23. 1901. COLORADO: West Spanish Peak, 6 Jul 1900, *Rydberg & Vreeland 5415* (HT: NY! on 2 sheets; IT: RM!).

E. laetevirens BTBC **28:** 506. 1901. WYOMING: Little Goose Creek, Jul 1899, *Tweedy 2005* (HT: NY!; IT: YU!).

E. leucotrichus BTBC **28:** 23. 1901. WYOMING. **Sheridan Co.:** Big Horn Mts., Jul 1899, *Tweedy 2003* (HT: NY!; IT: YU!).

E. mancus Fl. Rocky Mts. 902, 1067. 1917 [1918]. UTAH: La Sal Mts., 7 Jul 1911, *Rydberg & Garrett 8671* (HT: NY!; IT: RM!, US!, UT!).

E. nematophyllus BTBC **32:** 124. 1905. COLORADO. **Larimer Co.:** Near Dale Creek, 27 May 1897, *Osterhout s.n.* (HT: NY!; CT: US!).

E. oblanceolatus BTBC **24:** 294. 1897. MONTANA: Deer Lodge, 10 Jul 1895, *Rydberg 2822* (HT: NY!—designated by Rydberg in herb.; IT: NY!, US!).

E. peasei BTBC **32:** 126. 1905. COLORADO: Lake City, 10 Jun 1878, *Pease s.n.* (HT: NY!; IT: NY! dated 1878 only).

E. pulvinatum Fl. Rocky Mts. 911, 1067. 1917 [1918]. UTAH: Deep Creek, 6 Jun 1891, *Jones s.n.* (HT: US!; IT: BRY, CAS! no data, NY!, POM, RM!).

E. salicinus BTBC **32:** 125. 1905. COLORADO: Pagosa Springs, 31 Jul 1899, *Baker 670* (HT: NY!; IT: RM! 2 sheets).

E. smithii BTBC **32:** 125. 1905. COLORADO. **Gunnison Co.:** Parlin, 16 Aug 1901, *B. H. Smith 118* (HT: NY!).

E. spathulifolius BTBC **26:** 545. 1899. WYOMING: Black Rock Creek, Teton Forest Reserve, Aug 1897, *Tweedy 543* (HT: NY!; IT: YU!).

E. superbus E. L. Greene ex Rydb., Fl. Colorado 359, 361, 364. 1906. COLORADO: Near Pagosa Peak, 15 Aug 1899, *Baker 679* (LT: NY! designated by Rydberg in herb.; ILT: NDG, US!).

E. vetensis BTBC **32:** 126. 1905. COLORADO: Mountains near Veta Pass, 20 Jun 1900, *Rydberg & Vreeland 5427* (HT: NY!; IT: NEB, RM! dated 9 May).

E. viscidus BTBC **28:** 24. 1901. COLORADO: Gray-Back Mining Camps and Placer Gulch, 25–27 Jun 1900, *Rydberg & Vreeland 5416* (HT: NY!).

E. vreelandii BTBC **32:** 125. 1905. COLORADO: Wahatoya Creek, 7 Jul 1900, *Rydberg & Vreeland 5414* (HT: NY!; IT: RM!).

E. yukonensis Bull. New York Bot. Gard. **2:** 185. 1900. CANADA. YUKON: Dawson, 23 Jul 1899, *Williams s.n.* (HT: NY!).

Eriophyllum biternatum NAF **34:** 96. 1915. CALIFORNIA. **Tulare Co.:** Kern River, 2 Aug 1904, *Culbertson 4452* (HT: NY!; IT: CAS!).

E. bolanderi NAF **34:** 91. 1915. CALIFORNIA: Bear Valley, 1872, *Bolander & Keller s.n.* (HT: NY!; IT: NY!).

E. brachylepis NAF **34:** 88. 1915. CALIFORNIA: San Bernardino Mts., Bear Valley, 23 Jun 1894, *Parish 3344* (HT: NY!; IT: NY!).

E. cheiranthoides NAF **34:** 95. 1915. CALIFORNIA: Santa Barbara, Mission Hills, 21 Apr 1908, *Eastwood 21* (HT: NY!).

E. chrysanthum NAF **34:** 89. 1915. **California. Plumas Co.:** Quincy, 13 Jul 1913 [1893 in publication], *Heller 10855* (HT: NY!; IT: DS!).

E. cineraria NAF **34:** 93. 1915. **California. Plumas Co.:** Little Grizzly Creek below Genessee, 15 Jul 1907, *Heller & Kennedy 8849* (HT: NY!; IT: DS!).

E. crucigerum NAF **34:** 96. 1915. **MEXICO. Lower California:** Cedros Island, 18–20 Mar 1889, *Palmer 700* (HT: US!; IT: CAS! dated Mar 1889).

E. harfordii NAF **34:** 90. 1915. **Oregon:** The Dalles, 6 Jun 1869 [1889 in publication], *Harford & Dunn 453* (HT: NY!).

E. latilobum NAF **34:** 94. 1915. **California:** Back of San Mateo on the Half Moon Bay road, 28 May 1907, *Heller 8584* (HT: GH!; IT: DS!, NY!).

E. lutescens NAF **34:** 87. 1915. **California. Madera Co.:** Shuteye Mt., 20 Aug 1907, *Murdoch 2585* (HT: GH!; IT: NY!).

E. monoense NAF **34:** 87. 1915. **California:** Near summit of volcanic hills south of Mono Lake, 8 Jul 1863, *Brewer 1823* (HT: US!; IT: YU!).

E. superbum NAF **34:** 93. 1915. **Washington:** Quiniault, 2 Jul 1902, *Conrad 243* (HT: NY!).

E. trichocarpum NAF **34:** 89. 1915. **Nevada:** Palisades, 14 Jun 1882, *Jones s.n.* (HT: NY!; IT: CAS!, DS! 2 sheets, POM!, RM!).

E. tridactylum NAF **34:** 96. 1915. **California:** Yosemite Valley, Jul 1871, *Edwards s.n.* (HT: NY!).

Eupatorium holzingeri Brittonia **1:** 97. 1931. **Minnesota. Minona Co.:** Aug 1926, *Holzinger s.n.* (HT: NY!).

Euthamia glutinosa Brittonia **1:** 102. 1931. **Kansas:** Black Hills southeast of Belvidere, 15 Sep 1897, *Ward s.n.* (HT: NY!; IT: US!).

Flaveria pubescens NAF **34:** 145. 1915. **MEXICO. San Luis Potosí:** Río Verde, 2–8 Jun 1904, *Palmer 26* (HT: NY!; IT: NY!, US!).

Franseria caudata NAF **33:** 29. 1922. **New Mexico. Lincoln Co.:** Gray, 1 Sep 1898, *Skehan 108* (HT: NY!; IT: NY!, US!).

F. hispidissima NAF **33:** 29. 1922. **MEXICO. Zacatecas:** Real de Piños, *Diquet s.n.* (HT: NY!).

F. incana NAF 33: 30. 1922. **Arizona:** Ft. Huachuca, Sep 1892, *Wilcox s.n.* (HT: NY!; IT: NY!).

F. intricata NAF 33: 33. 1922. **MEXICO. Lower California:** San Bartólome Bay, 13 Mar 1911, *Rose 16196* (HT: NY!; IT: US!).

F. lancifolia NAF 33: 36. 1922. **MEXICO. Lower California:** Cedros Island, Mar–Jun 1897, *Anthony 290* (HT: US!).

F. malvacea NAF 33: 34. 1922. **MEXICO. [Sinaloa]:** Culiacán, 25 Oct–18 Nov 1891, *Palmer 1770* (HT: NY!; IT: US!).

F. palmeri NAF **33:** 25. 1922. **California:** San Diego, 1875, *Palmer s.n.* (HT: US!).

F. pringlei NAF **33:** 28. 1922. **MEXICO.** Coahuila Valley, Jimulco, 17 May 1885, *Pringle 192* (HT: NY!; IT: US!).

F. sanctae-gertrudis NAF **33:** 35. 1922. **MEXICO. Lower California:** Santa Gertrudis, Jan–Mar 1898, *Purpus 116* (HT: US!; CT: NY!).

F. strigulosa NAF **33:** 28. 1922. **California. Los Angeles Co.:** Sep 1899, *Chamberlain s.n.* (HT: NY!).

F. villosa NAF **33:** 26. 1922. **California. Mendocino Co.:** Mendocino, Jun 1898, *Brown 843* (HT: NY!; IT: US!).

Gaertneria linearis BTBC **32:** 133. 1905. **Colorado:** Calhan, Jul 1893, *Saunders s.n.* (HT: NY!).

Gaillardia crassa NAF **34:** 138. 1915. **Arizona:** Santa Rita Mts., 12 May 1884, *Pringle s.n.* (HT: NY!; IT: CAS!, NY!, US!).

G. crinita NAF **34:** 139. 1915. **Arizona:** Tucson, 4 Mar 1901, *Griffiths 2386* (HT: NY!).

G. flava NAF **34:** 139. 1915. **Utah:** Lower crossing, 2 Jul 1898, *Jones 6412* (HT: US!; IT: BRY, POM).

G. hallii NAF **34:** 135. 1915. **Oregon:** 1871, *Hall 279* (HT: NY!).

G. linearis NAF **34:** 137. 1915. **MEXICO. Chihuahua:** Near White Water, 18 Jun 1892, *Mearns 358* (HT: NY!; IT: DS!, US!).

G. mearnsii BTBC **37:** 443. 1910. **Arizona:** Fort Verde, 4 May 1888, *Mearns 322* (HT: NY! on 2 sheets).

G. nervosa NAF 34: 138. 1915. **MEXICO. Coahuila:** Carneros Pass, 11 Sep 1889, *Pringle 2309* (HT: NY!; IT: NY!, US!).

G. pringlei NAF **34:** 139. 1915. **Arizona:** Mesas near Tucson, Apr 1884, *Pringle s.n.* (HT: NY! on 2 sheets; IT: US!).

G. rigida Small ex Rydb., NAF **34:** 135. 1915. **Arkansas:** Fort Smith, 30 Jun 1853, *Bigelow 219* (HT: NY! on 2 sheets).

G. villosa NAF **34:** 135. 1915. **Texas. Bexar Co.:** *Jermy 40* (HT: US! no date; IT: NY! dated May 1904 and not numbered).

Gnaphalium lagopodioides Mem. New York Bot. Gard. **1:** 416. 1900. **Wyoming:** Yellowstone Park, Lower Geyser Basin, 4 Aug 1897, *Rydberg & Bessey 5134* (HT: NY!; IT: CAN, MONT, NEB, US!).

G. sulphurescens Mem. New York Bot. Gard. **1:** 415. 1900. **Wyoming:** Yellowstone Park, Lower Geyser Basin, 4 Aug 1897, *Rydberg & Bessey 5135* (ST: CAN, MONT, NEB, NY!, RM!, US! 2 sheets); Hot Springs, 1884, *Tweedy 172*; Mud Springs, 1871, Hayden Survey.

G. williamsii BTBC **37:** 324. 1910. **Montana:** Columbia Falls, 11 Aug 1894, *Williams s.n.* (HT: NY!; IT: MONT numbered *102D*).

Grindelia aphanactis BTBC **31:** 647. 1904 [1905]. **Colorado:** Durango, 21 Jul 1898, *Baker, Earle & Tracy 526* (HT: NY!; IT: GH!, MO! 2 sheets, NDG, NEB!, NY!, RM! 2 sheets, US!).

G. integerrima BTBC **37:** 128. 1910. **Idaho. Kootenai Co.:** Near Granite Station, 29 Jul 1892, *Sandberg, MacDougal & Heller 784* (HT: NY!; IT: DS!, NY!, US!).

G. laciniata Fl. Rocky Mts. 848, 1066. 1917 [1918]. **Utah:** Montezuma Canyon, east of Monticello, 13 Aug 1911, *Rydberg & Garrett 9692* (HT: NY!; IT: GH!, RM!, US!).

G. serrulata BTBC **31:** 646. 1904 [1905]. **Colorado:** Fort Collins, 10 Aug 1891, *Cowan s.n.* (HT: NY!).

Gutierrezia linearis BTBC **31:** 647. 1904 [1905]. **New Mexico. Lincoln Co.:** Gray, 2 Sep 1900, *Earle & Earle 474* (HT: NY!; IT: MO, TEX, US! dated Jul 1900).

G. scoparia BTBC **31:** 647. 1904 [1905]. **Colorado:**

Manitou, 7 Sep 1901, *Clements & Clements 16* (HT: NY!; IT: GH, MO).

Gymnolomia linearis BTBC **37:** 327. 1910. UTAH: St. George, 1877, *Palmer 241* (HT: NY!; IT: NY! 3 sheets).

Helenium centrale NAF **34:** 125. 1915. **GUATEMALA.** Cuajiniquilapa, May 1892, *Heyde & Lux 3403* (HT: NY!; IT: US!).

H. linifolium NAF **34:** 125. 1915. TEXAS: Kingsville, 11 Apr 1905, *Tracy 8949* (HT: NY!; IT: GH, MO, NEB, TEX, US!, WIS).

H. stenopterum NAF **34:** 123. 1915. **MEXICO.** SAN LUIS POTOSÍ: Media Luna near Río Verde, 2–8 Jun 1904, *Palmer 64* (HT: US!; IT: GH, MO, NY! 2 sheets).

Helianthus aridus BTBC **32:** 127. 1905. MONTANA: Great Falls, 20 Aug 1885, *Anderson 4217* (HT: NY!).

H. subrhomboideus Mem. New York Bot. Gard. **1:** 419. 1900. NEBRASKA. **Grant Co.:** 3 miles northeast of Whitman, 29 Jul 1893, *Rydberg 1627* (HT: NY! on 2 sheets; IT: NEB, US!).

Hieracium columbianum BTBC **28:** 513. 1901. IDAHO: Priest River Valley, 24 Jul 1900, *MacDougal 109* (HT: NY!; IT: RM!).

H. griseum Mem. New York Bot. Gard. **1:** 464. 1900. MONTANA: Jack Creek, 14 Jul 1897, *Rydberg & Bessey 5311* (LT: NY!—Cronquist, Univ. Wash. Publ. Biol. **17(5):** 237. 1955).

Hulsea carnosa Mem. New York Bot. Gard. **1:** 423. 1900. MONTANA: Mountains near Indian Creek, 22 Jul 1897, *Rydberg & Bessey 5194* (LT: NY!—Rydberg, NAF **34:** 40. 1914; ILT: MONT, RM!, US! 2 sheets).

H. mexicana NAF **34:** 41. 1914. **MEXICO.** San Pedro Martín Mts., 20 Jul 1905, *Goldman 1256* (HT: US!).

Hydropectis NAF **34:** 216. 1916. T: *Pectis aquatica* S. Wats.

Hymenoclea pentalepis NAF **33:** 14. 1922. ARIZONA. **Pima Co.:** 10 Apr 1901, *Griffiths 2630* (HT: NY!).

Hymenopappus cinereus BTBC **27:** 634. 1900. COLORADO: Walsenburg, 5 Jun 1900, *Rydberg & Vreeland 5479* (HT: NY!; IT: RM! numbered *4479*).

H. columbianus NAF **34:** 52. 1914. WASHINGTON. **Douglas Co.:** Near Moses Lake, 4 Jul 1893, *Sandberg & Leiberg 375* (HT: US!; IT: CAS!, NY!, RM! not numbered).

H. macroglottis BTBC **27:** 636. 1900. ARIZONA: Oak Creek, 5 Aug 1883, *Rusby s.n.* (HT: NY!; IT: NY!, US! both numbered *692*).

H. nanus NAF **34:** 53. 1914. NEVADA. **Elko Co.:** Above Cave Creek Post Office, 20 Aug 1908, *Heller 9494* (HT: NY!; IT: RENO!, US!).

H. niveus NAF **34:** 52. 1914. UTAH: Springdale, 14 May 1894, *Jones 5261* (HT: NY!; IT: US! dated 17 Mar).

H. petaloideus NAF **34:** 54. 1914. ARIZONA: Chiricahua Mts., Rock Creek Canyon, 6 Oct 1907, *Blumer 1202* (HT: NY!).

H. scaposus BTBC **27:** 634. 1900. ARIZONA: Vicinity of Flagstaff, 18 Jun 1898, *MacDougal 129* (HT: NY!; IT: RM!, US!).

H. sulphureus Rydb. ex Britton, Man. fl. n. states 1007. 1901. KANSAS. **Riley Co.:** 23 May 1895, *Norton 285* (HT: NY!; IT: RM!, US!).

H. tomentosus BTBC **27:** 633. 1900. UTAH: St. George, 1877, *Palmer 270* (HT: NY!; IT: BRY, ISC).

Hymenoxys quinquesquamata NAF **34:** 114. 1915. ARIZONA: Huachuca Mts., Carr Peak, 27 Aug 1910, *Goodding 874* (HT: NY!).

Isopappus validus Brittonia **1:** 100. 1931. KANSAS: Sandhills south of Arkansas River, about 6 miles south of Ellinwood, 23 Jul 1929, *Rydberg & Imler 1309* (HT: NY!).

Lactuca multifida Mem. New York Bot. Gard. **1:** 484. 1900. MONTANA: Columbia Falls, 25 Jul 1895, *Williams s.n.* (HT: NY!).

L. polyphylla BTBC **38:** 23. 1911. IDAHO: Lake Pend d'Oreille, 9 Aug 1885, *Greene s.n.* (HT: NY!).

Laphamia rotundata NAF **34:** 25. 1914. TEXAS: San Estaban, 3 Sep 1852, *Bigelow s.n.* (HT: NY!).

Lebetina cubana NAF **34:** 163. 1915. **CUBA.** Havana, 3 Dec 1904, *Curtiss 548* (HT: NY!; IT: US!).

Leptopharynx NAF **34:** 21. 1914. T: *Perityle parryi* A. Gray.

L. cordifolia NAF **34:** 22. 1914. **MEXICO.** SINALOA: Topolobampo, 23 Mar 1910, *Rose, Standley & Russell 13294* (HT: NY!; IT: GH!, US!).

L. lobata NAF **34:** 23. 1914. **MEXICO.** LOWER CALIFORNIA: Coronado, 19 Feb 1889, *Brandegee sn.* (HT: GH!; IT: UC, US!).

L. trisecta NAF **34:** 23. 1914. **MEXICO.** CHIHUAHUA: Near Chihuahua, 20 March 1885, *Pringle 183* (HT: NY!; IT: DS!, GH, UC, US!).

Leucactinia NAF **34:** 180. 1915. T: *Pectis bracteata* S. Wats.

Leuciva NAF **33:** 8. 1922. T: *Iva dealbata* A. Gray.

Liabellum NAF **34:** 294. 1927. T: *Liabum palmeri* A. Gray.

Liabum domingense NAF **34:** 291. 1927. [**DOMINICAN REPUBLIC**]. Constanza, Jun 1910, *von Türckheim 3113* (HT: NY!; IT: US!).

L. subacaule NAF **34:** 290. 1917. **HAITI.** Petit Borgne to Mt. Casse, 16 Aug 1903, *Nash 502* (HT: NY!).

Liatris bebbiana Brittonia **1:** 99. 1931. ILLINOIS: Fountaindale, *Bebb s.n.* (HT: NY!).

L. glabrata Brittonia **1:** 98. 1931. NEBRASKA. **Thomas Co.:** Plummer Ford, Dismal River, 23 Aug 1893, *Rydberg 1505* (HT: NY not found; IT: GH not found, US!).

L. haywardii Brittonia **1:** 99. 1931. SOUTH DAKOTA: Spearfish Canyon, 2 Aug 1926, *Hayward 166* (HT: NY!).

L. hirsuta Brittonia **1:** 98. 1931. MISSOURI. **Newton Co.:** Shoal Creek, near Reddings Mill, 14 Jul 1927, *Palmer 32505* (HT: NY!).

L. rosendahlii Brittonia **1:** 100. 1931. MINNESOTA: East of La Salle Springs, Itasca State Park, 17 Aug 1918, *Rosendahl 3699* (HT: MIN—fide Ownbey in litt. 1984).

Loxothysanus pedunculatus NAF **34:** 33. 1914. **MEXICO.** SAN LUIS POTOSÍ: Tamasopo Canyon, 24 Jun 1890, *Pringle 3096* (HT: NY!; IT: CAS!, US!).

Machaeranthera fremontii BTBC **32:** 123. 1905. COLORADO: Platte waters, 20 Jul 1844, *Fremont 421* (HT: NY!).

M. glabella E. L. Greene ex Rydb., Fl. Colorado 356–358. 1906. **Colorado:** Cerro Summit, 1 Aug 1901, *Baker 701* (LT: NY!—Turner, Phytologia **62:** 244. 1987; ILT: NDG, POM, RM, UC, US).

M. leptophylla BTBC **37:** 147. 1910. **Utah:** Logan, 9 Aug 1895, *Rydberg s.n.* (HT: NY!).

M. linearis Mem. New York Bot. Gard. **1:** 398. 1900. **Wyoming:** Yellowstone Park, 6 Aug 1885, *Letterman s.n.* (HT: NY!).

M. rubricaulis BTBC **28:** 596. 1901. **Colorado:** La Veta, 26 Sep 1900, *Vreeland 681* (HT: NY!; IT: NY!).

M. selbyi BTBC **32:** 123. 1905. **Colorado:** Southeast of Ouray, 7 Sep 1901, *Underwood & Selby 93a* (HT: NY!).

M. viscosula BTBC **32:** 124. 1905. **Colorado:** Veta Pass, 15 Jul 1896, *Shear 3655* (HT: NY!—designated by Rydberg in herb.).

Macronema grindelifolium Mem. New York Bot. Gard. **1:** 384. 1900. **Wyoming:** Yellowstone Park, Electric Peak, 18 Aug 1897, *Rydberg & Bessey 5045* (HT: NY!; IT: CAN, MONT, US!).

M. lineare Mem. New York Bot. Gard. **1:** 384. 1900. **Wyoming:** Gros Ventre River, 15 Jul–15 Aug 1897, *Tweedy 557* (HT: YU!).

M. obovatum BTBC 27: 618. 1900. **Utah:** City Creek Canyon, 25 Jul 1879, *Jones 1081* (HT: NY!; IT: CAS!, POM, RM, US, UTC).

M. obtusum BTBC **27:** 619. 1900. **Colorado. Chaffee Co.:** South Cottonwood Gulch, 14 Jul 1892, *Sheldon 5838* (HT: NY!).

Megaliabum NAF **34:** 293. 1927. T: *Vernonia andrieuxii* DC.

Mesadenia angustifolia BTBC **51:** 378. 1924. **Florida:** Everglades, Fort Lauderdale to Miami, Feb 1911, *Small, Carter & Small 3318* (HT: NY!).

Nesothamnus NAF **34:** 12. 1914. T: *Perityle incana* A. Gray.

Neurolaena intermedia NAF **34:** 307. 1927. **GUATEMALA. Alta Verapaz:** Cobán, Feb 1908, *von Türckheim 2105* (HT: NY!; IT: NY!, US!).

Nicolletia trifida NAF **34:** 180. 1915. **MEXICO. Lower California:** Los Angeles Bay, 1887, *Palmer 569* (HT: NY!; IT: NY!, US! numbered *369*).

Odontotrichum amplum BTBC **51:** 417. 1924. **MEXICO. Jalisco:** Between Heujuguilla and Mesquitec, 25 Aug 1897, *Rose 3575* (HT: US! on 2 sheets; IT: NY!).

O. cervinum BTBC **51:** 420. 1924. **MEXICO. [Nayarit]:** Tepic, between Pedro Paulo and San Blascito, 4 Aug 1897, *Rose 1986* (HT: US!; IT: NY!).

O. delphinifolium BTBC **51:** 419. 1924. **MEXICO. Zacatecas:** Sierra Madre, 18 Aug 1897, *Rose 2390* (HT: US!; IT: NY! dated Aug 1897).

O. filicifolium BTBC **51:** 414. 1924. **MEXICO. [Nayarit]:** Tepic, between Santa Gertrudis and Santa Teresa, 8 Aug 1897, *Rose 2101* (HT: US!; IT: NY!).

O. nephrophyllum BTBC **51:** 418. 1924. **MEXICO. Durango:** 13 Aug 1897, *Rose 2248* (HT: US!; IT: NY!).

O. scabrum BTBC **51:** 416. 1924. **MEXICO. Durango:** Sierra Madre, 13 Aug 1897, *Rose 3471* (HT: US!; IT: NY!).

Oligoneuron canescens BTBC **31:** 652. 1904 [1905]. **Wyoming:** Buffalo, Sep 1900, *Tweedy 3117* (HT: NY!).

Oreochrysum BTBC **33:** 152. 1906. T: *Haplopappus parryi* A. Gray.

Othake canescens NAF **34:** 60. 1914. **MEXICO. Nuevo León:** Monterey, 4 Jul 1888, *Pringle 1919* (HT: NY!).

O. macrolepis BTBC **37:** 332. 1910. **Colorado. Bent Co.:** Rule Creek, 18 Aug 1909, *Osterhout 4097* (HT: NY!).

O. robustum NAF **34:** 60. 1914. **MEXICO. Tamaulipas:** Tampico, 4 Jul 1896, *Pringle 6354* (HT: NY!; IT: CAS!).

Pectis biaristata NAF **34:** 211. 1916. **MEXICO. Sinaloa:** Fuerte, 26 Mar 1910, *Rose, Standley & Russell 13537* (HT: NY!; IT: US).

P. juniperina NAF **34:** 208. 1916. **CUBA. Pinar del Río:** 16 Dec 1911, *Shafer 11013* (HT: NY!; IT: US).

P. leonis NAF **34:** 207. 1916. **CUBA.** Santa Clara, North of St. Spiritus, Aug 1915, *León 5379* (HT: NY!).

P. mearnsii NAF **34:** 209. 1916. **Arizona:** Fort Verde, 1 Oct 1887, *Mearns 184* (HT: NY!).

P. multisetosa NAF **34:** 198. 1916. **GUATEMALA.** Chupadero, Santa Rosa, Nov 1892, *Heyde & Lux 4232* (HT: NY!; IT: US).

P. oerstediana NAF **34:** 213. 1916. **NICARAGUA. Chinandega:** Realejo, 19 Jan 1903, *Baker 2137* (HT: US!; IT: NY!).

P. polyantha NAF **34:** 207. 1916. **NICARAGUA. Chinandega:** Asseradores Island, 19 Jan 1903, *Baker 2133* (HT: GH!).

Perityle hofmeisteria NAF **34:** 18. 1914. **MEXICO. Durango:** Durango, Apr–Nov 1896, *Palmer 28* (HT: NY!; IT: GH, US!).

P. lineariloba NAF **34:** 18. 1914. **MEXICO. Durango:** San Ramón, 21 Apr–18 May 1906, *Palmer 89* (HT: NY!; IT: GH, UC, US!).

P. marginata NAF **34:** 14. 1914. **MEXICO. Lower California:** San José del Cabo, 25 Mar 1911, *Rose 16459* (HT: NY!; IT: GH, US!).

P. robusta NAF **34:** 16. 1914. **MEXICO. Lower California:** Cerralvo, 19 Apr 1911, *Rose 16880* (HT: NY!; IT: US!).

P. urticifolia NAF **34:** 15. 1914. **MEXICO. [Nayarit]:** Tepic, 5 Jan–6 Feb 1892, *Palmer 1960* (HT: NY!; IT: GH, UC, US!).

Petasites alaskanus NAF **34:** 314. 1927. **Alaska:** St. Paul's Island, 1879, *White s.n.* (HT: NY!).

P. hyperboreus NAF **34:** 312. 1927. **CANADA. British Columbia:** Second summit west of Shagit River, 22 Jul 1905, *Macoun* (herbarium *#76944*) *s.n.* (HT: NY!; IT: CAN).

Picradenia helenioides BTBC **28:** 21. 1901. **Colorado:** Sangre de Cristo Creek, 2 Jul 1900, *Rydberg & Vreeland 5495* (HT: NY!; IT: RM!).

Picradeniopsis Rydb. ex Britton, Man. fl. n. states 1008. 1901. T: *Bahia oppositifolia* Nutt.

Porophyllum cedrense Rose & Standley ex Rydb., NAF **34:** 189. 1916. **MEXICO. Lower California:** Cedros Island, 10 Mar 1911, *Rose 16099* (HT: US!; IT: NY!).

P. divaricatum NAF **34:** 186. 1916. **MEXICO. Sina-**

LOA: Mazatlán, 14 Apr 1899, *Goldman 386* (HT: NY!).

P. fruticulosum NAF **34:** 193. 1916. **MEXICO.** **COAHUILA:** Saltillo, 10–20 Nov 1902, *Palmer 285* (HT: NY!; IT: US!).

P. guatemalense NAF **34:** 190. 1916. **GUATEMALA.** Newton, 13–15 Dec 1895, *E. W. Nelson 3523* (HT: US!).

P. oblongum NAF **34:** 186. 1916. **MEXICO. [SINALOA]:** Culiacán, 25 Oct–18 Nov 1891, *Palmer 1807* (HT: NY!; IT: DS!, US!).

P. ochroleucum NAF **34:** 189. 1916. **MEXICO. LOWER CALIFORNIA:** Sancito, 4 Oct 1893, *Brandegee s.n.* (HT: NY!).

P. pinifolium NAF **34:** 192. 1916. **MEXICO. SONORA:** Alamos, 13 Mar 1910, *Rose, Standley & Russell 12757* (HT: NY!; IT: US!).

P. pittieri NAF **34:** 188. 1916. **GUATEMALA. BAJA VERAPAZ:** Cuesta de Quililhá, near Purulhá, 21 Apr 1905, *Pittier 146* (HT: US!).

P. porfyreum Rose & Standley ex Rydb., NAF **34:** 191. 1916. **MEXICO. LOWER CALIFORNIA:** San José del Cabo, 26 Mar 1911, *Rose 16478* (HT: US!).

Prenanthella BTBC **33:** 160. 1906. T: *Lygodesmia exigua* A. Gray.

Psacalium argutum BTBC **51:** 373. 1924. **MEXICO. SAN LUIS POTOSÍ:** 1879, *Schaffner 725* (HT: NY!).

P. langlassei BTBC **51:** 376. 1924. **MEXICO. GUERRERO:** Sierra Madre, 9 Jun 1898, *Langlassé 576* (HT: US!).

P. nelsonii BTBC **51:** 374. 1924. **MEXICO. OAXACA:** Vicinity of Cerro San Filipe, 1894, *E. W. Nelson 1111* (HT: US!).

Psathyrotopsis NAF **34:** 360. 1927. T: *Psathyrotes purpusii* T. S. Brandegee.

Pseudobartlettia NAF **34:** 358. 1927. T: *Psathyrotes scaposa* A. Gray.

Pseudoclappia J. Wash. Acad. Sci. **13:** 288. 1923. T: *Pseudoclappia arenaria* Rydb.

P. arenaria J. Wash. Acad. Sci. **13:** 289. 1923. **NEW MEXICO. Otero Co.:** White Sands, 20 Jul 1901, *Wooton s.n.* (HT: US!).

Psilostrophe divaricata NAF **34:** 8. 1914. **ARIZONA:** Grand Canyon of the Colorado, Aug 1897, *Allen s.n.* (HT: NY!).

P. grandiflora NAF **34:** 8. 1914. **ARIZONA:** Paradise, near Cedar Gulch, 21 Sep 1907, *Blumer 1709* (HT: NY!; IT: NEB, US!).

P. hartmanii NAF **34:** 8. 1914. **MEXICO. CHIHUAHUA:** Laguna de Guzmán, 16 Jul 1891, *Hartman 726* (HT: NY!; IT: US!).

P. villosa Rydb. ex Britton, Man. fl. n. states 1006. 1901. **KANSAS:** Typified to Kansas by Rydberg, NAF **34:** 7, 1914, which leaves 5 syntypes, all at NY!. Glass Mts., 13 Jul 1899, *White s.n.*; **Barber Co.:** Gypsum hills, 1896, *Hitchcock 741*; **Maude? Co.:** 26 Jun 1888, *Kellerman s.n.*; 1890, *Smyth s.n.*; **Clark Co.:** 14 Sep 1898, *White s.n.*

Ptilocalais macrolepis BTBC 38: 11. 1911. **UTAH:** Near Salt Lake City, 30 Apr 1904, *Garrett 182* (HT: NY!; IT: BRY, US! dated 2 May 1903).

Ptiloria ramosa Mem. New York Bot. Gard. **1:** 453. 1900. **NEBRASKA:** Bad lands north of Scott's Bluffs, 22 Jul 1891, *Rydberg 219* (HT: NY! on 2 sheets; IT: NEB 4 sheets).

Pyrrocoma acuminata BTBC **27:** 626. 1900. **WYOMING:** Fort Bridger, 1873, *Carter s.n.* (HT: NY!).

P. clementis BTBC **27:** 625. 1900. **COLORADO:** Mt. Harvard, 1896, *Clements 44* (HT: NY!).

P. erythropappa BTBC **27:** 624. 1900. **IDAHO:** Clear Water, *Spalding s.n.* (HT: NY!).

P. integrifolia var. *pumila* BTBC **27:** 626. 1900. **MONTANA:** Butte, 31 Jul 1895, *Rydberg 2808* (HT: NY! on 2 sheets).

P. lagopus BTBC **32:** 130. 1905. **WYOMING:** Hams Fork and La Barge, 13–21 Jul 1900, *Curtis s.n.* (HT: NY!).

P. rigida BTBC **27:** 624. 1900. **IDAHO. Kootenai Co.:** Granite Station, 29 Jul 1892, *Sandberg, MacDougal & Heller 785* (HT: NY! on 2 sheets; IT: DS!, US!).

P. villosa BTBC **27:** 625. 1900. **WYOMING. Sheridan Co.:** Big Horn Mts., Willow Creek, Aug 1899, *Tweedy 2063* (HT: NY!).

Raillardella minima NAF **34:** 319. 1927. **CALIFORNIA:** San Bernadino Mts., top of Grayback, 21 Aug 1907, *V. Bailey s.n.* (HT: US!; CT: NY!).

Raillardiopsis NAF **34:** 319. 1927. T: *Raillardella scabrida* Eastw.

Sartwellia puberula NAF **34:** 142. 1915. **TEXAS:** San Carlos, Rio Grande, *Parry 640* (HT: NY not found).

Schistocarpha kellermanii NAF **34:** 306. 1927. **GUATEMALA.** Volcán Agua, 4 Feb 1908, *Kellerman 7293* (HT: NY!).

S. longiligula NAF **34:** 305. 1927. **GUATEMALA.** San Miguel Uspanlón, Apr 1892, *Heyde & Lux 3383* (HT: GH!).

S. seleri NAF **34:** 305. 1927. **MEXICO. CHIAPAS:** Between San Martín and Ococingo, 13 Mar 1896, *Seler 2191* (HT: NY!).

S. velutina NAF **34:** 304. 1927. **MEXICO. VERACRUZ:** Mirador, 1842–1843, *Liebmann 490* (HT: GH!).

Senecio acutidens BTBC **27:** 180. 1900. **WYOMING:** Union Pass, 10 Aug 1894, *Nelson 858* (HT: NY!; IT: MO, RM!, US! dated 11 Aug).

S. alpicola Mem. New York Bot. Gard. **1:** 447. 1900. **MONTANA:** Cedar Mt., 16 Jul 1897, *Rydberg & Bessey 5269* (LT: NY!—Barkley, NAF II, **10:** 88. 1978).

S. altus Mem. New York Bot. Gard. **1:** 443. 1900. **MONTANA. Gallatin Co.:** Spanish Basin, 1 Jul 1897, *Rydberg & Bessey* 5258 (HT: NY!; IT: MONT, NEB, 2 sheets US!).

S. ambrosioides BTBC **37:** 467. 1910. **COLORADO. El Paso Co.:** Green Mt. Falls, 2 Aug 1892, *Sheldon s.n.* (HT: NY!).

S. arachnoides Mem. New York Bot. Gard. **1:** 442. 1900. **MONTANA:** Deer Lodge, 29 May 1891, *Kelsey s.n.* (HT: NY!; IT: NY!).

S. aurellus BTBC **27:** 182. 1900. **COLORADO:** Mancos, 24 Jun 1898, *Baker, Earle & Tracy 998* (HT: NY!; IT: RM!, US!).

S. canovirens BTBC **27:** 187. 1900. **NEW MEXICO. Lincoln Co.:** White Mts., 30 Jul 1897, *Wooton 244* (HT: NY!; IT: RM!, US!).

S. fedifolius BTBC **27:** 183. 1900. **COLORADO:** South Park, Aug 1871, *Canby s.n.* (HT: NY!).

S. flavovirens BTBC **27:** 181. 1900. **IDAHO:** Beaver Canyon, 28 Jun 1895, *Rydberg s.n.* (HT: NY!).

S. flintii BTBC **33:** 157. 1906. **Colorado:** Southwestern Colorado, *Flint s.n.* (HT: NY!).

S. fulgens BTBC **27:** 177. 1900. **Wyoming:** Grand Creek, Teton Forest Reserve, Aug 1897, *Tweedy 584* (HT: NY!; IT: YU!).

S. glaucescens Mem. New York Bot. Gard. **1:** 442. 1900. **Montana:** Yogo Baldy, 24 Aug 1896, *Flodman 913* (HT: NY! on 2 sheets).

S. glauciifolius Mem. New York Bot. Gard. **1:** 448. 1900. **Montana:** Columbia Falls, 1894, *Mrs. J. J. Kennedy 36* (HT: MONT—fide Plantenberg in litt. 1985).

S. greggii BTBC **27:** 170. 1900. **MEXICO. [Baja California]:** Santa Rosalía, 21 Apr 1897, *Gregg s.n.* (HT: NY!).

S. harbourii BTBC **33:** 158. 1906. **Colorado. Boulder Co.:** Mountains south of Ward, 18 Jul 1901, *Osterhout 2424* (HT: NY!; IT: RM! dated 17 Jul).

S. heterodoxus E. L. Greene ex Rydb., BTBC **27:** 178. 1900. **Colorado:** Little Kate Mine, *Baker, Earle & Tracy 569* (HT: NY!; IT: NEB).

S. hydrophiloides Mem. New York Bot. Gard. **1:** 441. 1900. **Idaho. Nez Perces Co.:** 29 Jul 1896, *Heller & Heller 3474* (HT: NY!; IT: DS).

S. idahoensis BTBC **27:** 183. 1900. **Idaho. Kootenai Co.:** Granite Station, 1892, *Sandberg, MacDougal & Heller 803* (HT: NY!; IT: CAS!).

S. jonesii BTBC **27:** 179. 1900. **Utah:** Alta, Wasatch Mts., 31 Jul 1879, *Jones 1125* (HT: NY! on 2 sheets; IT: POM, UTC!).

S. kingii BTBC **37:** 468. 1910. **Utah:** Cottonwood Canyon, Aug 1869, *Watson 676* (HT: NY!; IT: US!).

S. kuntzei Fl. Rocky Mts. 990, 1068. 1917 [1918]. **Colorado:** Sep 1874, *Kuntze 3033* (HT: NY!).

S. latus Mem. New York Bot. Gard. **1:** 444. 1900. **Montana:** Columbia Falls, *Mrs. J. J. Kennedy 28* (HT: MONT—fide Plantenberg in litt. 1985).

S. leonardii BTBC **37:** 468. 1910. **Utah:** Head of American Fork Canyon, 29 Jul 1885, *Leonard 143* (HT: NY!).

S. longipetiolatus BTBC **27:** 176. 1900. **Wyoming:** Spread Creek, Teton Forest Reserve, Aug 1897, *Tweedy 585* (HT: NY!; IT: YU! dated 15 Jul–15 Aug).

S. millelobatus BTBC **27:** 171. 1900. **New Mexico:** "Hills on the Limpia," 1851–1852, *Wright 1287* (HT: NY!; IT: NY!, US!).

S. multicapitatus Greenman ex Rydb., BTBC **33:** 160. 1906. **Colorado:** Huerfano Valley near Gardner, Sep 1900, *Vreeland 651* (HT: NY!—designated by Rydberg in herb.).

S. nelsonii Rydb. *in* A. Nels., BTBC **26:** 483. 1899. **Wyoming:** Centennial Hills, 9 Jun 1895, *Nelson 1297* (HT: NY!; IT: CAS!, RM!, UC, US!).

S. nephrophyllus Mem. New York Bot. Gard. **1:** 446. 1900. **Montana:** Big Blackfoot River, 1883, *Canby 203* (HT: NY not found).

S. oblanceolatus BTBC **27:** 175. 1900. **Colorado:** Como, South Park, 1 Aug 1895, *Crandall s.n.* (HT: NY!; IT?: CS no collector).

S. occidentalis var. *rotundatus* Mem. New York Bot. Gard. **1:** 438. 1900. **Idaho:** Mt. Chauvet, 29 Jul 1897, *Rydberg & Bessey 5268* (LT: NY!—Barkley, NAF II, 10: 128. 1978).

S. oödes BTBC **33:** 158. 1906. **Colorado:** Mt. Harvard, 1896, *Clements 39* (HT: NY!; IT: NEB 2 sheets).

S. pereziifolius Mem. New York Bot. Gard. **1:** 440. 1900. **Montana:** Jack Creek Canyon, 15 Jul 1897, *Rydberg & Bessey 5252* (HT: NY!; IT: US!).

S. platylobus BTBC **27:** 181. 1900. **Utah:** Wasatch Mts., May 1896, *Watson 671* (HT: NY!; IT: GH!, US!).

S. pseudaureus BTBC **24:** 298. 1897. **Montana:** Little Belt Mts., 10 Aug 1896, *Flodman 918* (HT: NY!; IT: NY!, US!)

S. rosulatus BTBC **27:** 188. 1900. **Colorado:** Georgetown, Jul 1885, *Patterson 79* (HT: NY!; IT: NY!, US!).

S. salicinus BTBC **27:** 186. 1900. **Colorado. Laramie [Larimer] Co.:** 3 Jun 1895, *Cowan s.n.* (HT: NY!).

S. saliens BTBC 24: 298. 1897. **Montana:** Little Belt Mts., Yogo Baldy, 24 Aug 1896, *Flodman 919* (LT: NY!—Barkley, NAF II, **10:** 131. 1978).

S. sanguisorboides BTBC **27:** 170. 1900. **New Mexico:** Santa Fe Canyon, 9 miles east of Santa Fe, 7 Jul 1897, *Heller & Heller 3820* (HT: NY!; IT: US!).

S. scribneri Mem. New York Bot. Gard. **1:** 441. 1900. **Montana:** Livingston, 5 Jun 1883, *Scribner 123b* (HT: NY!; IT: GH, US!).

S. semiamplexicaulis Mem. New York Bot. Gard. **1:** 440. 1900. **Wyoming:** Yellowstone Park, east DeLacy's Creek, 18 Aug 1897, *Rydberg & Bessey 5251* (HT: NY!; IT: MONT, NEB, RM!, US! dated 10 Aug).

S. solidago Mem. New York Bot. Gard. **1:** 439. 1900. **Montana:** Tiger Butte, 31 Jul 1886, *Williams 264* (ST: MONT); Alhambra, 5 Sep 1892, *Kelsey s.n.* (ST: MONT).

S. solitarius Mem. New York Bot. Gard. **1:** 444. 1900. **Wyoming:** Yellowstone Park, Aug 1885, *Tweedy 713* [*813* in publication] (HT: YU!; IT: NY!).

S. subcuneatus BTBC **27:** 179. 1900. **Colorado:** Grizzly Creek, 19 Jul 1896, *Baker s.n.* (HT: NY!).

S. tracyi BTBC **33:** 159. 1906. **Colorado:** Bob Creek, west of Mt. Hesperus, 1 Jul 1898, *Baker, Earle & Tracy 276* (HT: NY!; IT: CS, NY!, RM!).

S. tridenticulatus BTBC **27:** 175. 1900. **Colorado. Chaffee Co.:** Cottonwood Creek, Buena Vista, 4 Jul 1892, *Sheldon s.n.* (HT: NY!; IT: COLO numbered *5832*).

S. turbinatus BTBC **33:** 159. 1906. **Colorado:** Lake City, 3 May 1878, *Pease s.n.* (HT: NY!).

S. tweedyi BTBC **37:** 469. 1910. **Wyoming:** Buffalo Fork, Aug 1897, *Tweedy 586* (HT: NY!).

S. variifolius Mem. New York Bot. Gard. **1:** 438. 1900. **Montana:** Clendennin, 12 Jul 1882, *Williams 205* (LT: MONT—Barkley, NAF II, **10:** 131. 1978, fide Plantenberg in litt. 1985).

Sideranthus glaberrimus BTBC **27:** 621. 1900. **Nebraska:** Sand Hills, central Nebraska, 27 June 1893, *Rydberg 1403a* (HT: NY!).

S. puberulus BTBC **27:** 622. 1900. **Colorado:** Salida, *Clarke 174* (HT: NY!)

S. turbinellus BTBC **27:** 622. 1900. **Idaho:** Pocatello, 25 Aug 1892, *Mulford s.n.* (HT: NY!).

Sinclairia brachypus NAF **34:** 299. 1927. **GUATEMALA.** Volcán Ipala, Jan 1907, *Pittier 1886* (HT: US!; CT: NY!).
S. pittieri NAF **34:** 300. 1927. **COSTA RICA.** Alajuelita, Dec 1894, *Pittier 9093* (HT: US!; CT: NY!).
Sinclairiopsis NAF **34:** 292. 1927. T: *Liabum klattii* B. L. Robinson & Greenman.
Solidago canadensis var. *gilvocanescens* Contr. U.S. Natl. Herb. **3:** 162. 1895. **NEBRASKA:** Cody's Lakes at the head of Dismal River, 10 Aug 1893, *Rydberg 1662* (HT: US! designated by Rydberg in herb.; IT: NY! 2 sheets).
S. garrettii BTBC **37:** 134. 1910. **UTAH. Salt Lake Co.:** Big Cottonwood Canyon, 28 Aug 1906, *Garrett 2041* (HT: NY!; IT: UT!).
S. glaucophylla BTBC **37:** 133. 1910. **MONTANA:** Dry Plains near Kalispel, Flathead Valley, 25 Jul 1901, *MacDougal 760* (HT: NY!; IT: NY! 2 sheets).
S. hapemaniana Brittonia **1:** 101. 1931. **NEBRASKA. Kearney Co.:** 13 Jul 1900, *Rydberg 5402* (HT: NY! on 2 sheets).
S. laevicaulis BTBC **31:** 649. 1904 [1905]. **COLORADO:** North Park, 3 Sep 1899, *Osterhout 4* (HT: NY!; IT: RM! numbered *1797*).
S. lunellii Brittonia **1:** 101. 1931. **NORTH DAKOTA. Benson Co.:** Leeds, 13 Aug 1909, *Lunell s.n.* (HT: NY!).
S. nivea BTBC **37:** 133. 1910. **MONTANA:** Lima, Aug 1895, *Rydberg 2804* (HT: NY!).
S. polyphylla BTBC **31:** 651. 1904 [1905]. **COLORADO:** Canyon City, 1896, *Clements 295* (HT: NY!; IT: NEB).
S. radulina BTBC **31:** 650. 1904 [1905]. **UTAH:** Cottonwood Canyon, Aug 1869, *Watson 558* (HT: NY!).
S. rubra BTBC **31:** 649. 1904 [1905]. **COLORADO:** North Park, 3 Sep 1899, *Osterhout 5* (HT: NY!).
S. serra BTBC **31:** 650. 1904 [1905]. **COLORADO:** Yampa, 4 Aug 1898, *Shear & Bessey 5274* (HT: NY!).
S. somesii Brittonia **1:** 101. 1931. **IOWA:** Elmira, 24 Aug 1909, *Somes 3687* (HT: NY!).
S. viscidula BTBC **31:** 651. 1904 [1905]. **COLORADO:** Grand Lake, 2 Aug 1888, *Holway s.n.* (HT: NY!).
Stenotopsis BTBC **27:** 617. 1900. Ts: *Haplopappus linearifolius* DC., *H. interior* Coville.
Stenotus andersonii BTBC **27:** 615. 1900. **MONTANA:** Belt Mts., 14 Jul 1886, *Anderson 3561* (HT: NY!).
S. borealis Rydb. *in* Britton & Rydb., Bull. New York Bot. Gard. **2:** 184. 1901. **CANADA. YUKON:** Foot of Lake Lebarge, 23 Jun 1899, *Tarleton 51* (HT: NY!; IT: US!).
S. falcatus BTBC **27:** 616. 1900. **UTAH:** Red Creek, 1877, *Palmer 212* [*202* in publication] (HT: NY!; IT: ISC, US! numbered *202* and not dated).
Tagetes alamensis NAF **34:** 154. 1915. **MEXICO. SONORA:** Sierra de Alamos, 19 Mar 1910, *Rose, Standley & Russell 13083* (HT: NY!; IT: US!).
T. heterocarpha NAF **34:** 155. 1915. **MEXICO. JALISCO:** Near Guadalajara, 4 Oct 1889, *Pringle 2488* (HT: NY!; IT: NY!, US!).
T. seleri NAF **34:** 151. 1915. **MEXICO. CHIAPAS:** Comitán, 14 Aug 1896, *Seler 3085* (HT: NY!).
Taraxacum alaskanum BTBC **28:** 512. 1901. **ALASKA:** Point Barrow, 23 Jul 1898, *McIlhenny s.n.* (HT: NY!; IT: CAN, RM! numbered *111*).
T. eriophorum Mem. New York Bot. Gard. **1:** 454. 1900. **MONTANA:** Sheridan, 5–8 May 1892, *Fitch 15* (HT: MONT—fide Plantenberg in litt. 1985).
T. leiospermum BTBC **32:** 137. 1905. **COLORADO:** Tennessee Pass, 18 Jul 1902, *Osterhout 2645* (HT: NY not found; IT: RM!).
Tetracarpum flavum NAF **34:** 46. 1914. **MEXICO. OAXACA:** Sierra de San Felipe, 10 Oct 1894, *Smith 263 & 626* (HT: NY!; IT: US!, both numbers on one sheet with one label).
T. guatemalense NAF **34:** 45. 1914. **GUATEMALA:** Llano de San Juan de Dios, Oct 1865, *Bernoulli 135* (HT: NY!).
Tetradymia linearis BTBC **32:** 130. 1905. **UTAH:** Rock Creek, 1877, *Palmer 264* (HT: NY!).
Tetraneuris angustifolia BTBC **32:** 128. 1905. **NEW MEXICO. Lincoln Co.:** White Mts., 21 Aug 1891, *Wooton 374* (HT: NY!; IT: NY!; RM!, US!).
T. crandallii BTBC **32:** 127. 1905. **COLORADO:** Grand Junction, 28 May 1894, *Crandall s.n.* (HT: NY!).
T. pilosa E. L. Greene ex Rydb., Fl. Colorado 378, 379. 1906. **COLORADO:** Los Pinos (Bayfield), May 1899, *Baker 726* (HT: NY!; IT: US!).
T. septentrionalis BTBC **37:** 447. 1910. **IDAHO. Palouse Co.:** Jun–Jul 1892, *Aiton s.n.* (HT: NY!; IT: NY!).
T. stenophylla Rydb. ex Britton, Man. fl. n. states 1009, 1901. **KANSAS. Ford Co.:** May 1891, *Ellis s.n.* (HT: NY!).
Thelesperma intermedium BTBC **27:** 631. 1900. **NEBRASKA. Banner Co.:** Aug 1890, *Rydberg 192* (HT: NY!; IT: NEB).
T. marginatum Mem. New York Bot. Gard. **1:** 421. 1900. **CANADA.** Assiniboia, Medicine Hat, 31 May 1894, *Macoun 5073* (HT: NY!; IT: US!).
T. tenue BTBC **27:** 631. 1900. **COLORADO:** Veta Pass, 3 Jul 1900, *Rydberg & Vreeland 5473* (HT: NY!).
Thymophylla canescens NAF **34:** 178. 1915. **MEXICO. SAN LUIS POTOSÍ:** 22°N Lat, 1878, *Parry & Palmer 515* (HT: NY!; IT: F, GH, US!).
T. gracilis NAF **34:** 176. 1915. **TEXAS:** Barstow, Jul 1901, *Earle 622* (HT: NY!)
T. myriophylla NAF **34:** 176. 1915. **TEXAS:** Laredo, 21 Mar 1903, *Reverchon 3982* (HT: NY!; IT: SMU).
T. pringlei NAF **34:** 177. 1915. **MEXICO. CHIHUAHUA:** Chihuahua, 5 Oct 1886, *Pringle 1022* (HT: NY!; IT: DS!, NY!).
T. puberula NAF **34:** 177. 1915. **MEXICO. SAN LUIS POTOSÍ:** 1877, *Schaffner 328* (HT: NY!; IT: F, GH, NY! 2 sheets, US! dated 1879).
T. villosula NAF **34:** 177. 1915. **TEXAS:** [Western Texas in publication] *Wright s.n.* (HT: NY! no data).
Townsendia intermedia Rydb. ex Britton, Man. fl. n states 944. 1901. **KANSAS. Trego Co.:** Prairie, 1896, *Rich 718* (HT: NY!; IT: GH!, MO, RM!, US!).
T. vreelandii BTBC **28:** 22. 1901. **COLORADO:** Side of Veta Mt., 19 Jul 1900, *Vreeland 639* (HT: NY!; IT: CAN, RM!).
Trichaetolepis NAF **34:** 170. 1915. T: *Adenophyllum wrightii* A. Gray.
Trichymenia NAF **34:** 56. 1914. T: *Hymenothrix wrightii* A. Gray.
Venegazia deltoidea NAF **34:** 5. 1914. **MEXICO.**

Lower California: Sangal, 7 Apr 1886, *Orcutt s.n.* (HT: NY!; IT: US!).

Vesicarpa NAF **34:** 242. 1916. T: *Artemisia potentilloides* A. Gray.

Xylorrhiza brandegei BTBC **32:** 124. 1905. **Colorado:** San Juan Pass of Sierra Madre, Sep 1875, *Brandegee s.n.* (HT: GH not found; IT: CS, NY! numbered *1165*).

X. lanceolata BTBC **37:** 146. 1910. **Southern Utah, Northern Arizona:** [Utah, St. George in publication], 1877, *Palmer 288* (HT: NY! on 2 sheets; IT: US!).

BALSAMINACEAE

Impatiens aurella BTBC **28:** 34. 1901. **Idaho:** Priest River Valley, MacAbee's Ranch, 23 Jul 1900, *MacDougal 20* (HT: NY!).

I. mexicana NAF **25:** 95. 1910. **MEXICO. [Veracruz]:** Orizaba, 1853, *Müller s.n.* (HT: NY!)

I. nortonii NAF **25:** 95. 1910. **Kansas. Pottawatomie Co.:** 23 Jul 1895, *Norton 58* (HT: NY!; IT: US!).

I. occidentalis NAF **25:** 94. 1910. **Washington:** North fork of Nooksack River, 28 Jul 1890, *Suksdorf 960* (HT: NY!; IT: US!).

BETULACEAE

Betula crenata Rydb. ex Butler, BTBC **36:** 429. 1909. **Montana:** Rost Lake, 15–16 Jul 1901, *MacDougal 665* (HT: NY!).

B. subcordata Rydb. ex Butler, BTBC **36:** 436. 1909. **Idaho. Nez Perces Co.:** Hatwai Creek, 24 Apr 1892, *Sandberg, MacDougal & Heller 33* (HT: NY!; IT: US!).

BORAGINACEAE

Amsinckia rugosa Fl. Rocky Mts. 729, 1066. 1917 [1918]. **Idaho. Nez Perces Co.:** Lewiston, 27 Apr 1896, *Heller & Heller 2975* (HT: NY!).

Cryptantha confusa BTBC **36:** 679. 1909. **Wyoming:** Yellowstone Park, upper Madison Canyon, 3 Aug 1897, *Rydberg & Bessey 4884* (HT: NY!; IT: MONT, NEB 2 sheets, US!).

C. grandiflora BTBC **36:** 679. 1909. **Idaho. Nez Perces Co.:** Valley of Clearwater River, 23 Apr 1892, *Sandberg, MacDougal & Heller 10* (HT: NY!; IT: CAS!, DS).

C. leptophylla BTBC **36:** 678. 1909. **Utah:** St. George, 1877, *Palmer 350* (HT: NY!).

C. minima BTBC **28:** 31. 1901. **Colorado:** Cuchara River above La Veta, 31 May 1900, *Rydberg & Vreeland 5697* (HT: NY!).

Eremocarya muricata BTBC **36:** 677. 1909. **Utah:** Southern Utah, 1874, *Parry 164* (HT: NY not found; IT: NY!, ISC).

Eritrichium aretioides var. *elongatum* Mem. New York Bot. Gard. **1:** 327. 1900. **Montana. Gallatin Co.:** Spanish Basin, 26 Jun 1897, *Rydberg & Bessey 4891* (HT: NY!—designated by Rydberg in herb.).

Heliotropium spathulatum BTBC **30:** 262. 1903. **Montana:** Great Falls, 16 Jul 1890, *Williams 542* (HT: NY!; IT: MONT dated 11 Jul, US! dated 22 Jun 1889).

Lappula angustata BTBC **31:** 636. 1904 [1905]. **South Dakota:** Oreville, 16 Jul 1892, *Rydberg 891* (HT: NY!; IT: NY! 2 sheets, US!).

L. besseyi BTBC **31:** 636. 1904 [1905]. **Colorado:** Cheyenne Canyon, neighborhood of Pike's Peak, 25 Jul 1895, *Bessey s.n.* (HT: NY!).

L. calycosa BTBC **28:** 30. 1901. **Colorado:** Walsenburg, 5 Jun 1900, *Rydberg & Vreeland 5715* (HT: NY!; IT: RM!).

L. coerulescens Mem. New York Bot. Gard. **1:** 328. 1900. **Montana:** Bridger Mts., 18 Jun 1897, *Rydberg & Bessey 4899* (HT: NY!; IT: CAN, MONT, NEB, RM!, US!).

L. leptophylla Mem. New York Bot. Gard. **1:** 329. 1900. **Montana. Gallatin Co.:** Aug 1886, *Tweedy 1174* (ST: YU!); Bozeman, 10 Jul 1887, *Tweedy 223* (ST: YU!). **Wyoming:** Yellowstone Park, Mammoth Hot Springs, Jun 1885, *Tweedy 807* (ST: YU!). No type material of this species was seen by Gentry & Carr, Mem. New York Bot. Gard. **26:** 1976. Of the three above collections only *807* has *Lappula leptophylla* written in Rydberg's hand but he did not write TYPE on the sheet.

L. leucotricha BTBC **36:** 676. 1909. **Arizona:** Tucson, 20 Apr 1894, *Toumey s.n.* (HT: NY!).

Lithospermum lanceolatum Mem. New York Bot. Gard. **1:** 333. 1900. **Idaho. Nez Perces Co.:** Lewiston, 20 May 1896, *Heller & Heller 3092* (HT: NY!).

Mertensia alba BTBC **31:** 638. 1904 [1905]. **Colorado:** La Plata River, 13 Jun 1898, *Baker, Earle & Tracy 825* (HT: NY!; IT: NEB).

M. canescens BTBC **31:** 640. 1904 [1905]. **Colorado:** Berthoud Pass, Jul 1903, *Tweedy 5664* (HT: NY!; IT: RM!).

M. humilis BTBC **36:** 681. 1909. **Wyoming. Albany Co.:** Sand Creek, 2 Jun 1900, *Nelson 7043* (HT: NY!; IT: RM!).

M. intermedia Mem. New York Bot. Gard. **1:** 335. 1900. **Montana:** Bridger Mts., 18 Jun 1897, *Rydberg & Bessey 4873* (HT: NY!).

M. leonardii BTBC **36:** 680. 1909. **Utah:** Mill Creek Canyon, 31 Jul 1884, *Leonard s.n.* (HT: NY!).

M. lineariloba BTBC **28:** 32. 1901. **Colorado:** Trinchera Range, west Indian Creek, 14–15 Jun 1900, *Rydberg & Vreeland 5691* (HT: NY!).

M. membranacea BTBC **28:** 33. 1901. **Idaho:** Priest River Valley, MacAbee's Ranch, 23 Jul 1900, *MacDougal 3* (HT: NY!).

M. myosotifolia A. A. Heller ex Rydb., Fl. Colorado 289, 290, 292. 1906. **Colorado. Eagle Co.:** Red Cliff, 26 Jun 1900, *Osterhout 2164* (LT: NY!—Williams, Ann. Missouri Bot. Gard. **24:** 118. 1937; ILT: GH, MO, RM!, UC).

M. obtusiloba BTBC **28:** 32. 1901. **Colorado:** Pikes Peak, 1900, *Clements s.n.* (HT: NY!).

M. ovata BTBC **28:** 32. 1901. **Colorado:** West Spanish

Peak, 6 Jul 1900, *Rydberg & Vreeland 5690* (HT: NY!; IT: CS, NEB, RM!).

M. pallida BTBC **36:** 680. 1909. MONTANA: Lima, 28 Jun 1895, *Rydberg 2777* (HT: NY).

M. parryi BTBC **31:** 639. 1904 [1905]. COLORADO: Alpine ridges lying east of Middle Park, 1861, *Parry 286* (HT: NY!; IT: NY!).

M. perplexa BTBC **31:** 639. 1904 [1905]. COLORADO. **Boulder Co.:** Mts. south of Ward, 18 Jul 1901, *Osterhout 2439* (HT: NY!; IT: RM!).

M. picta BTBC **31:** 638. 1904 [1905]. COLORADO. **Larimer Co.:** Estes Park, 20 Jul 1903, *Osterhout 2823* (HT: NY!; IT: RM! labeled Moraine Pass).

M. polyphylla var. *platensis* BTBC **31:** 638. 1904 [1905]. COLORADO: La Plata Mts., Bob Creek, 26 Jun 1898, *Baker, Earle & Tracy 180* in part (HT: NY!; IT: NY!).

M. subpubescens BTBC **30:** 261. 1903. MONTANA. **Gallatin Co.:** Spanish Basin, 23 Jun 1897, *Rydberg & Bessey 4876* (HT: NY!; IT: MONT, NEB, RM!).

M. tubiflora BTBC **26:** 544. 1899. WYOMING: Big Horn Mts., headwaters of the Tongue River, Jul 1898, *Tweedy 119* (HT: NY!; IT: YU!).

M. tweedyi Mem. New York Bot. Gard. **1:** 336. 1900. MONTANA: Old Hollowtop near Pony, 7 & 9 Jul 1897, *Rydberg & Bessey 4867* (HT: NY!; IT: CAN, MONT, NEB, RM!).

M. viridula BTBC **31:** 639. 1904 [1905]. COLORADO: North Cheyenne Canyon, neighborhood of Pikes Peak, 16 Jul 1894, *Bessey s.n.* (HT: NY!; IT: NEB).

Oreocarya argentea BTBC **31:** 637. 1904 [1905]. COLORADO. **Garfield Co.:** Rifle, 23 Jun 1900, *Osterhout 2122* (HT: NY!).

O. eulophus BTBC **31:** 637. 1904 [1905]. COLORADO: Dolores, 16 Jun 1892, *Crandall s.n.* (HT: NY!).

O. macounii Eastw. ex Rydb., BTBC **40:** 480. 1913. CANADA. NORTHWEST TERRITORIES: Moose Mountain Creek, 6 Jul 1880, *Macoun s.n.* (HT: NY!).

O. pustulosa BTBC **40:** 480. 1913. UTAH: Elk Mts., Hammond Canyon, 31 Jul 1911, *Rydberg & Garrett 9320* (HT: NY!).

O. spicata BTBC **36:** 678. 1909. COLORADO: Artist's Glen, 1 Aug 1901, *Clements & Clements 102* (HT: NY!; IT: NEB, RM!).

BRASSICACEAE

Arabis brevisiliqua BTBC **39:** 326. 1912. CANADA. BRITISH COLUMBIA: Skagit Valley, 6 Jul 1905, *Macoun* (herbarium *#70825) s.n.* (HT: NY!; IT: CAN).

A. egglestonii Fl. Rocky Mts. 361, 1062. 1917 [1918]. COLORADO: Clover Mt. above Garfield, 31 Jul 1910, *Eggleston 6013* (HT: NY!—designated by Rydberg in herb.).

A. macdougalii BTBC **39:** 326. 1912. MONTANA: Old Sentinel, mountain near Missoula, 12 Jun 1901, *MacDougal 191* (HT: NY!; IT: CAN, NY!, US!).

A. oblanceolata BTBC **31:** 557. 1904. COLORADO: Valley Spur, 13 Sep 1901, *Underwood & Selby 454* (HT: NY!).

A. oreophila BTBC **34:** 437. 1907. UTAH: Divide between Big Cottonwood Canyon and Heber Valley, 5 Jul 1905, *Rydberg & Carlton 6678* (HT: NY!).

A. philonipha A. Nels. ex Rydb., Fl. Colorado 164, 165. 1906. WYOMING. **Albany Co.:** Telephone Mines, 1 Aug 1900, *Nelson 7913* (HT: NY! on 2 sheets; IT: GH).

A. selbyi BTBC **31:** 557. 1904. COLORADO: West of Ouray, 9 Sep 1901, *Underwood & Selby 207* (HT: NY!).

S. spatifolia Fl. Rocky Mts. 361, 1062. 1917 [1918]. COLORADO. **Larimer Co.:** Estes Park, 20 Jul 1903, *Osterhout 2808* (HT: NY!; IT: RM! 2 sheets labeled Moraine Pass).

A. stokesiae Fl. Rocky Mts. 361, 1062. 1917 [1918]. UTAH: Wasatch Mts., Parley's Canyon, 8 Jun 1901, *Stokes s.n.* (HT: NY!; IT: CAS!, US!).

Barbarea americana Mem. New York Bot. Gard. **1:** 174. 1900. MONTANA: Bridger Mts., 10 Jun 1897, *Rydberg & Bessey 4145* (HT: NY! on 3 sheets).

Cardamine cardiophylla BTBC **28:** 280. 1901. COLORADO. **Lake Co.:** Tennessee Pass, 27 Jun 1900, *Osterhout 2178* (HT: NY!; IT: RM!).

C. multifolia BTBC **29:** 238. 1902. IDAHO: Beaver Canyon, 26 Jun 1895, *Rydberg s.n.* (HT: NY!; IT: NY!).

C. pratensis L. var. *acaulis* Bot. Not. **1882:** 63. 1882. SWEDEN. VÄSTERGÖTLAND: Hjerpås, 22 Jun 1881, *Rydberg s.n.* (LT: LD—fide Almborn in litt. 1989; ILT: NY!).

C. unijuga BTBC **24:** 246. 1897. MONTANA: Madison Range, Spanish Basin, 18 Jul 1896, *Flodman 494* (HT: NY! on 2 sheets; IT: US!).

Cartiera cuneata Fl. Rocky Mts. ed. 2, 1123. 1922 [1923]. UTAH. **San Pete Co.:** Manti, 26 Jun 1920, *Garrett 2938* (HT: NY!).

Cheirinia brachycarpa BTBC **39:** 325. 1912. UTAH: Abajo Mts., 17–20 Aug 1911, *Rydberg & Garrett 9713* (HT: NY!; IT: RM! dated 17 Aug, US, UT! dated 17 Aug).

Chlorocrambe BTBC **34:** 435. 1907. T: *Caulanthus hastatus* S. Wats.

Draba argyrea BTBC **30:** 251. 1903. IDAHO: Sawtooth Mts., head of Pettit Lake, *Henderson 3538* (HT: US!; IT: NY!).

D. aureiformis BTBC **28:** 278. 1901. COLORADO: Headwaters of Pass Creek, 30 Jun 1900, *Rydberg & Vreeland 6157* (HT: NY!).

D. brachystylis BTBC **29:** 240. 1902. UTAH: Wasatch Mts., Alta, 7 Aug 1879, *Jones 1357* (HT: NY!; IT: CAS!, POM, RM!, US!).

D. cana BTBC **29:** 241. 1902. CANADA. ALBERTA: Morley, 12 Jun 1887, *Macoun s.n.* (HT: NY!; IT: CAN!).

D. coloradensis BTBC **31:** 555. 1904. COLORADO: Fort Collins, 1 May 1897, *Armstrong 109* (HT: NY!; IT: NY! 2 sheets, RM!, US!).

D. columbiana BTBC **29:** 242. 1902. CANADA. BRITISH COLUMBIA: Revelstoke, 23 May 1890, *Macoun s.n.* (HT: NY!; IT: US!).

D. crassa Mem. New York Bot. Gard. **1:** 182. 1900. COLORADO: Gray's Peak, 1895, *Rydberg s.n.* [Division of Agrostology] (HT: US? not found).

D. decumbens BTBC **29:** 240. 1902. COLORADO: Gray's Peak, 1895, *Rydberg s.n.* (HT: NY!).

D. macouniana BTBC **29:** 240. 1902. CANADA. BRIT-

ISH COLUMBIA: Summit of Rocky Mts., 15 Aug 1890, *Macoun s.n.* (HT: NY!).
D. mccallae BTBC **29:** 241. 1902. **CANADA. ALBERTA:** Valley below Mt. Aylmer, 4 Aug 1899, *McCalla 2267* (HT: NY!).
D. parryi BTBC **29:** 241. 1902. **COLORADO:** Foot of Gray's Peak, Aug 1872, *Greene s.n.* (HT: NY!).
D. sobolifera BTBC **30:** 251. 1903. **UTAH:** Marysvale, 28 Aug 1894, *Jones 5936* (HT: US!; IT: NY!, POM).
D. streptocarpa var. *grayana* BTBC **31:** 555. 1904. **COLORADO:** Gray's Peak, Aug 1895, *Rydberg s.n.* (HT: NY!).
D. uncialis BTBC **30:** 251. 1903. **UTAH:** Marysvale, Tate mine, 28 Aug 1894, *Jones 5940am* (HT: US!; IT: POM).
Erysimum angustatum Rydb. *in* Britton & Rydb., Bull. New York Bot. Gard. **2:** 171. 1901. **CANADA. YUKON:** Dawson, 13 Jul 1899, *Williams s.n.* (HT: NY!; IT: US).
E. oblanceolata BTBC **31:** 557. 1904. **COLORADO:** Georgetown, Aug 1895, *Rydberg s.n.* (HT: NY!).
E. radicatum BTBC **31:** 558. 1904. **COLORADO:** Bottomless Pit (Pike's Peak), 6 Jul 1901, *Clements & Clements 441* (HT: NY!; IT: NEB, RM!, US).
Lepidium brachybotryum BTBC 34: 427. 1907. **UTAH:** Juab, 10 Jun 1902, *Goodding 1075* (HT: NY!; IT: NEB, RM!, US!).
L. crandallii BTBC **34:** 427. 1907. **COLORADO:** Palisades, 14 May 1898, *Crandall 131* (HT: NY!; IT: CS 2 sheets, RM! 2 sheets one not numbered).
L. elongatum BTBC **29:** 234. 1902. **WASHINGTON. Whitman Co.:** Almata, 13 Jun 1896, *Elmer 21* (HT: NY!).
L. fletcheri BTBC **34:** 428. 1907. **CANADA. MANITOBA:** Winnipeg, 12 Jul 1905, *Fletcher s.n.* (HT: NY!).
L. georginum BTBC **30:** 253. 1903. **UTAH:** Southern Utah, 1874, *Parry 19* (HT: NY!; IT: ISC).
L. jonesii BTBC **29:** 233. 1902. **UTAH:** St. George, 7 Apr 1880, *Jones 1636* (HT: NY! on 2 sheets; IT: CAS!, POM, RM!, US!).
Lesquerella ovalifolia Rydb. ex Britton & Brown, Ill. Fl. Northern States and Canada **2:** 137. 1897. **NEBRASKA. Kimball Co.:** Hills of upper Lawrence Fork, 11 Aug 1891, *Rydberg 22* (HT: NY!; IT: US!).
L. shearis BTBC **29:** 237. 1902. **COLORADO:** Idaho Springs, 29 Aug 1895, *Shear 3269* (HT: NY!).
L. spatulata Contr. U.S. Natl. Herb. **3:** 486. 1896. **SOUTH DAKOTA:** Deadwood, 5 Jul 1892, *Rydberg 534* (HT: US!; IT: NEB, NY! 2 sheets).
L. utahensis BTBC **30:** 252. 1903. **UTAH:** American Fork Canyon, 31 Jul 1880, *Jones 1354* (HT: NY!; IT: CAS!, NY!, US!).
Parrya platycarpa BTBC **39:** 326. 1912. **Utah:** Uintah Mts., Aug 1869, *Watson 54* (HT: NY!; IT: US!).
Physaria acutifolia BTBC **28:** 279. 1901. **COLORADO:** Grand Junction, Jun 1892, *Eastwood s.n.* (HT: NY!; IT: DS!).
P. brassicoides BTBC **29:** 237. 1902. **NEBRASKA:** Canyon south of Scott's Bluff, 22 Jul 1891, *Rydberg 24* (HT: NY!; IT: NEB 2 sheets, RM!).
P. floribunda BTBC **28:** 279. 1901. **COLORADO:** Sangre de Cristo Creek, 2 Jul 1900, *Rydberg & Vreeland 6135* (HT: NY!; IT: RM!).
P. vitulifera BTBC **28:** 278. 1901. **COLORADO:** Idaho Springs, 26 Aug 1895, *Rydberg s.n.* (HT: NY!; IT: NY!, RM! dated just 1895).
Pilosella novae-angliae Torreya **7:** 158. 1907. **VERMONT:** Willoughby Mt., 16 Jul 1887, *Faxon s.n.* (HT: NY!; IT: NY!).
P. stenocarpa Torreya **7:** 161. 1907. **CANADA.** Assiniboia, Wood Mt., 6 Jun 1885, *Macoun* (herbarium *#10007*) *s.n.* (HT: NY!; IT: CAN).
Pleurophragma BTBC **34:** 433. 1907. Ts: *Pachypodium integrifolium* Nutt., *Pleurophragma platypodium* Rydb., *Thelypodium integrifolium* var. *gracilipes* B. L. Robinson.
P. platypodum BTBC **34:** 434. 1907. **UTAH:** Moab, 30 Aug 1891, *Jones s.n.* (HT: NY! on 2 sheets; IT: CAS!, DS!, RM!, US!).
Rorippa [*Roripa* in publication] *clavata* BTBC **29:** 235. 1902. **WASHINGTON. Chehalis Co.:** Hogeman, 29 Jun 1897, *Lamb 1221* (HT: NY!; IT: US!).
R. [*Roripa* in publication] *integra* BTBC **29:** 236. 1902. **Utah:** Wasatch Mts., Aug 1869, *Watson 64* in part (HT: NY!; IT: GH!, US!).
R. [*Roripa* in publication] *underwoodii* BTBC **29:** 235. 1902. **COLORADO:** Red Mt., south of Ouray, 11 Sep 1901, *Underwood 299a* (HT: NY!).
Schoenocrambe decumbens BTBC **31:** 409. 1904. **MONTANA:** Melrose, 6 Jul 1895, *Rydberg 2671* (HT: NY!; IT: NY!).
Smelowskia lineariloba BTBC **31:** 555. 1904. **COLORADO:** Douglass Mt., 18 Jul 1878, *Jones 447* (HT: NY! on 2 sheets).
S. lobata BTBC **39:** 327. 1912. **CANADA. ALBERTA:** Northern Rocky Mts., *Bourgeau s.n.* [Palliser Expedition] (HT: NY!).
Sophia glandulifera BTBC **28:** 281. 1901. **WYOMING:** Between Sheridan and Buffalo, 15 Jun–15 Jul 1900, *Tweedy 3592* (HT: NY!; IT: US!, YU!).
S. gracilis Mem. New York Bot. Gard. **1:** 475. 1900. **WYOMING:** Yellowstone Park, Jun 1888, *C.H. Hall s.n.* (HT: NY!).
S. intermedia Mem. New York Bot. Gard. **1:** 184. 1900. **MONTANA:** Melrose, 1895, *Rydberg 2672*; Bridger Mts., 12 Jun 1897, *Rydberg & Bessey 4199* (ST: NY! 2 sheets); Bridger Mts., 14 Jun 1897, *Rydberg & Bessey 4200* (ST: NY! 2 sheets); Cedar Mt., 16 Jul 1897, *Rydberg & Bessey 4198*; West Gallatin, 1883, *Scribner 8p*; **Beaver Head Co.:** Jun 1888, *Tweedy 116* (ST: YU!); Bozeman, 1892, *Shaw*; Helena, 1890, *Kelsey*; **Custer Co.:** 1892, *Light*; Yellowstone Park, 1888, *Hall*; east fork of Yellowstone, 1885, *Tweedy 575* (ST: YU! not numbered).
S. leptophylla BTBC **29:** 239. 1902. **COLORADO. Larimer Co.:** 20 Aug 1895, *Crandall & Cowan 55* (HT: NY!).
S. leptostylis BTBC **39:** 325. 1912. **UTAH:** Big Cottonwood Canyon, below Silver Lake, 4 Jul 1905, *Rydberg & Carlton 6629* (HT: NY! on 2 sheets; IT: RM!, US!).
S. magna BTBC **34:** 436. 1907. **COLORADO:** Bluffs north of La Veta, 21 May 1900, *Rydberg & Vreeland 6163* (HT: NY! on 2 sheets; IT: NEB, RM! 2 sheets).
S. millefolia Rydb. ex Britton, Man. fl. n. states ed. 2,

462. 1905. Kentucky River, May 1834, *Peter s.n.* (HT: NY!).

S. nelsonii BTBC **34:** 436. 1907. **Wyoming:** Yellowstone Park, Wraith Falls, 7 Jul 1899, *Nelson & Nelson 5710* (HT: NY!; IT: DS! 2 sheets, RM! 4 sheets, US!).

S. purpurascens BTBC **31:** 556. 1904. **Colorado:** Red Mt., south of Ouray, 11 Sep 1901, *Underwood & Selby 275* (HT: NY!).

S. ramosa BTBC **31:** 556. 1904. **Colorado:** Pitkin, 13 Sep 1901, *Underwood & Selby 413* (HT: NY!).

S. viscosa BTBC **29:** 238. 1902. **Idaho:** Beaver Canyon, 26 Jun 1895, *Shear 3029* (HT: NY!).

Stanleya arcuata BTBC **29:** 232. 1902. **Nevada:** Unionville Valley, Jun 1868, *Watson 105* in part (HT: NY!).

S. canescens BTBC **29:** 232. 1902. **Utah:** Frisco, 22 Jun 1880, *Jones 1809* (HT: NY!; IT: NY!, POM).

S. glauca BTBC **31:** 409. 1904. **Colorado:** Prairie, entrance Soldier's Canyon to Sulphur Springs, 15 Jul 1895, *Cowan s.n.* (HT: NY!; IT: CS, NY!, RM! numbered *99*).

S. runcinata BTBC **30:** 252. 1903. **Idaho?:** 1896, *Henderson s.n.* (HT, US!).

Stanleyella BTBC **34:** 435. 1907. T: *Thelypodium wrightii* A. Gray.

Streptanthella Fl. Rocky Mts. 364, 1062. 1917 [1918]. T: *Arabis longirostris* S. Wats.

Thelypodiopsis BTBC **34:** 432. 1907. T: *Thelypodium elegans* M. E. Jones.

Thelypodium leptosepalum BTBC **34:** 433. 1907. **Idaho. Nez Perces Co.:** Lewiston, 6 May 1896, *Heller & Heller 3022* (HT: NY!; IT: US!).

T. macropetalum BTBC **29:** 233. 1902. **Utah:** Farmington, 14 May 1881, *Jones 1841* (HT: NY!; IT: BRY, DS!, NY!, POM, US!, UTC!).

T. ovalifolium BTBC **30:** 253. 1903. **Utah:** Panguitch Lake, 7 Sep 1894, *Jones 6015e* (HT: US!; IT: POM; CT: NY!).

T. palmeri BTBC **34:** 432. 1907. **Southern Utah, Northern Arizona:** 1877, *Palmer 25* (HT: NY!; IT: BRY, ISC, NY! 2 sheets, US! 2 sheets).

T. utahense BTBC **29:** 233. 1902. **Utah:** St. George, 9 Apr 1880, *Jones 1648* (HT: NY!; IT: NY!, POM, US!, UT!).

Thlaspi coloradense BTBC **28:** 280. 1901. **Colorado:** Bald Mt., 1 Aug 1896, *Bessey s.n.* (HT: NY!).

T. purpurascens BTBC **28:** 281. 1901. **Arizona:** 1876, *Palmer 571* (HT: NY!; IT: NY! 2 sheets).

Thysanocarpus trichocarpus BTBC **30:** 253. 1903. **Utah:** Silver Reef, 4 May 1894, *Jones 5163b* in part (HT: US!; IT: NY! 2 sheets).

CAMPANULACEAE

Bolelia brachyantha Mem. New York Bot. Gard. **1:** 483. 1900. **Montana:** Augusta, 30 Jul 1887, *Williams 712* (HT: NY!; IT: MONT, US!).

Campanula macdougalii BTBC **28:** 25. 1901. **Idaho:** Priest River Valley, MacAbee's Ranch, 23 Jul 1900, *MacDougal 66* (HT: NY!).

Lobelia kalmii var. *strictiflora* Mem. New York Bot. Gard. **1:** 378,. 1900. **Montana:** Teton River, 24 Aug 1883, *Scribner 130* (LT: NY!—McVaugh, NAF **32A:** 67. 1942; ILT: PH).

CAPPARACEAE

Cleomella cornuta BTBC **30:** 249. 1903. **Utah:** Cainsville, 20 Jul 1894, *Jones 5656* (HT: US; IT: BRY, NY! 2 sheets, RM).

CAPRIFOLIACEAE

Linnaea serphyllifolia J. New York Bot. Gard. **8:** 135. 1907. **Alaska:** Cape Nome, 1900, *Blaisdell s.n.* (HT: NY!).

Lonicera ebractulata Mem. New York Bot. Gard. **1:** 372. 1900. **Montana:** Spanish Basin, 28 Jul 1897, *Rydberg & Bessey 5010* (HT: NY!).

L. hirsuta var. *glaucescens* Contr. U.S. Natl. Herb. **3:** 503. 1896. **South Dakota:** Piedmont and Little Elk Creek, 28 Jun 1892, *Rydberg 747* (HT: US!; IT: NEB).

Sambucus microbotrys BTBC **28:** 503. 1901. **Colorado:** Bottomless Pit, neighborhood of Pike's Peak, 11 Aug 1896, *Bessey s.n.* (HT: NY!).

Symphoricarpos parishii BTBC **26:** 545. 1899. **California. San Bernardino Co.:** San Bernardino Mts., Mill Creek, 3 Jul 1892, *Parish 2514* (LT: NY!—G. N. Jones, J. Arnold Arbor. **21:** 241. 1940; ILT: K).

S. utahensis BTBC **26:** 544. 1899. **Utah:** Logan, 9 Aug 1895, *Rydberg s.n.* (HT: NY!).

S. vaccinioides Mem. New York Bot. Gard. **1:** 371. 1900. **Montana:** Forks of the Madison, 24 Jul 1897, *Rydberg & Bessey 5017* (HT: NY!; IT: US!).

CARYOPHYLLACEAE

Alsine curtisii BTBC **28:** 274. 1901. **Wyoming:** Headwaters of Cliff Creek, 9–18 Aug 1900, *Curtis s.n.* (HT: NY!).

A. palmeri BTBC **39:** 315. 1912. **Utah:** Beaver Valley, 1877, *Palmer 54* (HT: NY!).

A. polygonoides E. L. Greene ex Rydb., Fl. Colorado 127, 128. 1906. **Colorado:** Near Pagosa Peak, Aug 1899, *Baker 307* (HT: NY!).

Alsinopsis macrantha BTBC **31:** 407. 1904. **Colorado:** La Plata Mts., Little Kate Basin, 14 Jul 1898, *Baker, Earle & Tracy 678* (HT: NY!; IT: NY!, RM! 2 sheets).

Arenaria cephaloidea BTBC **39:** 316. 1912. **Washington:** Spokane, 10 Sep 1902, *Kreager 617* (HT: NY!; IT: UTC!).

A. confusa BTBC **28:** 275. 1901. **New Mexico. Lincoln Co.:** White Mts., 6 Aug 1897, *Wooton 295* (HT: NY!; IT: DS!, US!).

A. eastwoodiae BTBC **31:** 406. 1904. **Colorado:** Grand Junction, May 1892, *Eastwood s.n.* (HT: NY!).

A. fendleri var. *porteri* BTBC **31:** 407. 1904. **Colorado:** Gray's Peak, 29 Aug 1878, *Jones 716* (HT: NY! on 2 sheets; IT:UTC!).

A. polycaulos BTBC **31:** 406. 1904. **Colorado:** Silverton, 3 Jul 1898, *collector unknown* (from Herb. State Agric. College, Colo.) *s.n.* (HT: NY; IT: RM! numbered *73*).

A. subcongesta var. *lithophila* Mem. New York Bot. Gard. **1:** 148. 1900. **Montana:** Madison Range, Spanish Peaks, 14 Jul 1896, *Flodman 437* (HT: NY!—designated by Rydberg in herb.).

A. tweedyi BTBC **31:** 406. 1904. **Colorado:** La Plata Mts., 15 Jul 1896, *Tweedy s.n.* (HT: NY!).

Cerastium earlei BTBC **30:** 249. 1903. **Colorado:** Near La Plata Post Office, 14 Jul 1898, *Baker, Earle & Tracy 472* (HT: NY!; IT: DS!, RM! 2 sheets, US!).

C. graminifolium BTBC **30:** 250. 1903. **Washington:** Pullman, Jun 1897, *Elmer 177* (HT: NY!).

C. leibergii BTBC **30:** 250. 1903. **Idaho:** Upper St. Marys River, 28 Jun 1895, *Leiberg 1103* (HT: US!; IT: NY!).

C. pilosum E. L. Greene ex Rydb., Fl. Colorado 128, 129. 1906. **Colorado:** Mountains above Ouray, 10 Aug 1901, *Baker 745* (HT: NY!; IT: RM! 2 sheets, US!).

C. pulchellum BTBC **31:** 405. 1904. **Colorado:** Hayden Peak, 14 Jul 1898, *Baker, Earle & Tracy 573* (HT: NY!).

C. thermale BTBC **39:** 314. 1912. **Wyoming:** Lower Geyser Basin, 4 Aug 1897, *Rydberg & Bessey 4025* (HT: NY! on two sheets).

Lychnis striata BTBC **31:** 408. 1904. **Colorado:** Cameron Pass, 30 Jul 1896, *Baker s.n.* (HT: NY!; IT: NY!, US!).

Paronychia wardii Rydb. ex Small, Fl. S.E. U.S. 400, 1330. 1903. **Oklahoma:** Alva, 14 Oct 1897, *Ward 62* (HT: NY!).

Silene antirrhina var. *depauperata* BTBC **31:** 408. 1904. **CANADA. British Columbia:** Lower Arrow Lake, 18 Jun 1890, *Macoun s.n.* (HT: NY!: IT: CAN).

S. antirrhina var. *vaccarifolia* BTBC **31:** 407. 1904. **Idaho. Nez Perces Co.:** Lewiston, 29 May 1896, *Heller & Heller 3153* (HT: NY!).

CHENOPODIACEAE

Atriplex buxifolia BTBC **39:** 311. 1912. **Wyoming. Sheridan Co.:** Dayton, Sep 1899, *Tweedy 2456* [*2656* in publication] (HT: NY!; IT: YU!).

A. garrettii BTBC **39:** 312. 1912. **Utah:** Moab and vicinity, 1–2 Jul 1911, *Rydberg & Garrett 8465* (HT: NY!; IT: US!, UT!).

A. lapathifolia Mem. New York Bot. Gard. **1:** 133. 1900. **Montana:** Fridley on the Yellowstone River, 1 Sep 1887, *Tweedy 278* (HT: YU!; IT: NY!).

A. oblanceolata BTBC **31:** 403. 1904. **Colorado:** Delta, 3 Sep 1897, *Cowan 4071* (HT: NY!; IT: CS 2 sheets, NEB, RM! 2 sheets, US! 2 sheets, YU! 2 sheets).

A. odontoptera BTBC **31:** 404. 1904. **Wyoming:** Buffalo, Sep 1900, *Tweedy 3302* (HT: NY!; IT: RM!, YU!).

A. subconferta Fl. Rocky Mts. 248, 1061. 1917 [1918]. **Idaho:** Twin Falls and Shoshone Falls, 27 Jul 1911, *Nelson & Macbride 1379* (HT: NY!; IT: DS!, RM!, US!).

Blitum hastatum BTBC **28:** 273. 1901. **Wyoming:** Buffalo, Sep 1900, *Tweedy 3295* (HT: NY!; IT: YU!).

Chenopodium atrovirens Mem. New York Bot. Gard. **1:** 131. 1900. **Montana.** Electric Peak, 18 Aug 1897, *Rydberg & Bessey 3948* (HT: NY!; IT: CAN, MONT, NEB).

C. pratericola BTBC **39:** 309. 1912. **Kansas. Riley Co.:** 2 Aug 1895, *Norton 436* (HT: NY!; IT: US!).

C. wolfii BTBC **30:** 248. 1903. **Colorado:** Twin Lakes, Aug 1873, *Wolf 253* (HT: NY!—designated by Rydberg in herb.; IT: US!).

Corispermum [*Coriospermum* in publication] *emarginatum* BTBC **31:** 404. 1904. **Wyoming:** Laramie, 15 Sep 1897, *Nelson 4282* (HT: NY!; IT: RM!).

C. marginale BTBC **30:** 247. 1903. **New Mexico:** Albuquerque, 1894, *Herrick s.n.* (HT: NY!; IT: US!).

C. [*Coriospermum* in publication] *villosum* BTBC **24:** 191. 1897. **Montana:** Manhattan, 17 Jul 1895, *Rydberg 2623* (HT: NY!—designated by Rydberg in herb; IT: NY!, US!).

Endolepis ovata BTBC **30:** 248. 1903. **Wyoming:** Buffalo, Sep 1900, *Tweedy 3290* (HT: NY!).

Eurotia subspinosa BTBC **39:** 312. 1912. **Utah:** St. George, 15 May 1902, *Goodding 810* (HT: NY!; IT: RM!, US!).

CONVOLVULACEAE

Cressa erecta BTBC **40:** 466. 1913. **Utah. Salt Lake Co.:** Near Beck's Hot Springs, Jul 1905, *Garrett 870f* (HT: NY!).

CORNACEAE

Cornella BTBC **33:** 147. 1906. Ts: *Cornus canadensis* L., *C. suecica* L., *C. unalaschkensis* Ledeb.

Suida interior BTBC **31:** 572. 1904. **Nebraska:** Dismal River south of Thedford, 27 Jun 1893, *Rydberg 1414* (HT: NY!; IT: NEB).

S. stolonifera var. *riparia* BTBC **31:** 573. 1904. **Colorado:** Crystal Creek, 27 Jun 1901, *Baker 257* (HT: NY!; IT: RM! 2 sheets).

CRASSULACEAE

Sedum frigidum BTBC **28:** 282. 1901. **Montana:** Old Hollow Top near Pony, 7 Jul 1897, *Rydberg & Bessey 4248* (HT: NY!; IT: MONT, NEB, RM!, US!).

S. polygamum BTBC **28:** 283. 1901. **Colorado:** West Spanish Peak, 9 Jul 1900, *Rydberg & Vreeland 6113* (HT: NY!).

CUSCUTACEAE

Cuscuta gracilis BTBC **28:** 501. 1901. **Wyoming:** Rolling plains between Sheridan and Buffalo, 15 Jun–15 Jul 1900, *Tweedy 3492* [*3292* in publication] (HT: NY!; IT: US!, YU!).

C. megalocarpa BTBC **28:** 501. 1901. **Colorado:** Cucharas Creek, near La Veta, 15 Sep 1900, *Vreeland 670* (HT: NY!—designated by Rydberg in herb.).

CYPERACEAE

Carex ebenea BTBC **28:** 266. 1901. **Colorado:** Pike's Peak, 1900, *Clements s.n.* (HT: NY!).

C. pseudoscirpoidea Mem. New York Bot. Gard. **1:** 78. 1900. **Montana:** Spanish Basin, Jul 1896, *Rydberg 3064* (LT: NY!—Mackenzie, NAF **18:** 209. 1935).

Eleocharis thermalis Mem. New York Bot. Gard. **1:** 69. 1900. **Montana:** Mud Springs, 1881, *Porter* [Hayden Survey] *s.n.* (ST: NY!). Rydberg stated "characters of the fruit are taken from Tweedy's specimen." **Wyoming:** Yellowstone Park, east fork of Firehole River, Sep 1884, *Tweedy 222* (ST: US!, YU!); Yellowstone Park, Lower Geyser Basin, 4 Aug 1897, *Rydberg & Bessey 3812* (ST: MONT, NY! 2 sheets, RM!).

ELATINACEAE

Elatine rubella Mem. New York Bot. Gard. **1:** 260. 1900. **Wyoming:** Yellowstone Park, Yellowstone Lake, Sep 1885, *Tweedy 439* (HT: YU!; IT: US!).

E. williamsii Mem. New York Bot. Gard. **1:** 260. 1900. **Montana:** Lower Sand Coulee, 1891, *Williams 855* (ST: MONT, NY! both dated 25 Aug, NY! dated Sep); Missoula, 1898, *Williams & Griffith s.n.*

ERICACEAE

Azaleastrum Mem. New York Bot. Gard. **1:** 297. 1900. T: *Rhododendron albiflorum* Hook.

Phyllodoce hybrida Mem. New York Bot. Gard. **1:** 298. 1900. **Montana:** Old Hollowtop near Pony, 7 Jul 1897, *Rydberg & Bessey 4657* (HT: NY! on 2 sheets; IT: US!).

Vaccinium globulare Mem. New York Bot. Gard. **1:** 300. 1900. **Montana:** Spanish Creek, Jul 1886, *Tweedy 1170* (ST: YU!); Prickly Pear Creek, 1883, *Scribner 133*; Jocko River, 1883, *Canby 217*.

EUPHORBIACEAE

Euphorbia albicaulis Mem. New York Bot. Gard. **1:** 266. 1900. **Nebraska. Cheyenne Co.:** South of Pottes, 13 Aug 1891, *Rydberg 356* (HT: NY!; IT: US!).

FABACEAE

Amorpha arizonica NAF **24:** 33. 1919. **Arizona:** Huachuca Mts., Ramsey Canyon, 10 Jun 1909, *Goodding 136* (HT: NY!; IT: NEB, NY!, RM!).

A. bushii NAF **24:** 31. 1919. **Florida:** Chattahoochee River, 12 Aug 1897, *Bush 13* (HT: NY!).

A. crenulata NAF **24:** 30. 1919. **Florida:** Between Cocoanut Grove and Cutler, 9 May 1904, *Small & Wilson 1898* (HT: NY!).

A. curtissii NAF **24:** 30. 1919. **Florida:** Jacksonville, 6 May & 21 Aug 1894, *Curtiss 4703* (HT: NY!; IT: US! dated 21 Aug).

A. floridana NAF **24:** 31. 1919. **Florida:** *Chapman s.n.* (HT: NY!).

Aragallus alpicola Mem. New York Bot. Gard. **1:** 252. 1900. **Montana:** Old Hollowtop near Pony, 9 Jul 1897, *Rydberg & Bessey 4503* (LT: NY!—Barneby, Proc. Calif. Acad. Sci. IV, **27:** 262. 1952; ILT: GH, MONT, NEB 2 sheets, NDG, RM!, US!).

A. angustatus BTBC **34:** 422. 1907. **Nebraska:** Rush Creek, 2 Jul 1891, *Rydberg 82c* (HT: NY!; IT: NEB, NY! dated 1 Jul).

A. argophyllus Mem. New York Bot. Gard. **1:** 255. 1900. **Montana:** Little Blackfeet River, 1860, *Cooper s.n.* (HT: NY!; IT: US!).

A. atropurpureus BTBC **34:** 424. 1907. **Wyoming:** Big Horn Mts., headwaters of Tongue River, Jul 1898, *Tweedy 125* (HT: NY!).

A. besseyi Mem. New York Bot. Gard. **1:** 250. 1900. **Montana. Gallatin Co.:** Spanish Basin, 23 Jun 1897, *Rydberg & Bessey 4501* (HT: NY! on 2 sheets; IT: NEB 2 sheets, RM!, US!).

A. patens BTBC **34:** 421. 1907. **Colorado:** Near Boulder, Jul 1902, *Tweedy 5164* (HT: NY!; IT: RM!).

A. varians Rydb. *in* Britton & Rydb., Bull. New York Bot. Gard. **2:** 176. 1901. **CANADA. Yukon:** Lewis River, 28 Jun 1899, *Tarleton 33b* (HT: NY!; IT: US!).

A. villosus BTBC **28:** 36. 1901. **Montana:** Craig, 20 Jun 1900, *Wilcox 378* (HT: US!; IT: NY!).

A. viscidulus Mem. New York Bot. Gard. **1:** 253. 1900. **Montana:** Melrose, 6 Jul 1895, *Rydberg 2716* (LT: NY!—Barneby, Proc. Calif. Acad. Sci. IV, **27:** 241. 1952).

A. viscidulus var. *depressus* Mem. New York Bot. Gard. **1:** 253. 1900. **Montana. Park Co.:** Haystack Peak, 6 Aug 1887, *Tweedy 120* (HT: YU!; IT: NY not found).

Astragalus brevidens NAF **24:** 450. 1929. **Oregon. Baker Co.:** 3 miles north of Whitney, 22 Jul [Feb in publication] 1921, *Peck 10360* (HT: NY!; IT: DS! 2 sheets).

A. cusickii BTBC **26:** 541. 1899. **Oregon:** Malheur, 1885, *Cusick 1238* (HT: GH!).

A. earlei E. L. Greene ex Rydb., NAF **24:** 444. 1929. **Texas:** Limpia Canyon, 15 Apr–3 May 1902 [5 Apr in publication], *Earle & Tracy 226* (HT: NY!; IT: GH!, US!).

A. halei NAF **24:** 447. 1929. **Louisiana:** *Hale 313* (HT: NY!).

A. hartmanii NAF **24:** 442. 1929. **MEXICO. Chihuahua:** San Diego Ranch, 2 May 1891, *Hartman 678* (HT: NY!; IT: GH, K, US!).

A. kelseyi Mem. New York Bot. Gard. **1:** 241. 1900. **Montana:** Deer Lodge, 9 Jun 1892, *Kelsey s.n.* (LT: NY!—Barneby, Mem. New York Bot. Gard. **13:** 531. 1964; ILT: MONT, MU, RSA).

A. macounii Mem. New York Bot. Gard. **1:** 243. 1900. **CANADA. British Columbia:** Deer Park, Lower Arrow Lake, 5 Jun 1890, *Macoun 25* (HT: NY!; IT: US! not numbered).

A. marcidus E. L. Greene ex Rydb., NAF **24:** 446. 1929.

TEXAS: Davis Mts., 28 Apr 1902, *Tracy & Earle 319* (HT: NY!; IT: G, GH, MO, NDG, TEX, US!).

A. oreophilus BTBC **31:** 561. 1904. COLORADO: Pagosa Springs, Jul 1899, *Baker 419* (HT: NY!; IT: MO, NDG, RM!, US!).

A. pervelutinus NAF **24:** 444. 1929. **MEXICO.** CHIHUAHUA: Chihuahua, 1 Apr 1885, *Pringle 189* (HT: NY!; IT: US!).

A. prunifer Mem. New York Bot. Gard. **1:** 239. 1900. **WYOMING. Carbon Co.:** Medicine Bow, 5 Jun 1897, *Nelson 3143* (HT: NY!; IT: PH).

A. sanguineus NAF **24:** 443. 1929. **MEXICO.** NUEVO LEÓN: Near Saltillo, Feb–Oct 1880, *Palmer 243* (HT: US!; IT: GH, K, P, POM).

A. shearii BTBC **31:** 562. 1904. COLORADO: Twin Lakes, 21 Aug 1896, *Shear 3317* (HT: NY!; IT: RM!, US!).

A. sulphurescens BTBC **28:** 36. 1901. COLORADO: Georgetown, 18 Aug 1895, *Rydberg s.n.* (HT: NY!).

A. tarletonis Rydb. *in* Britton & Rydb., Bull. New York Bot. Gard. **2:** 175. 1901. **CANADA.** YUKON: Five-finger Rapids, 5 Jul 1899, *Tarleton 78* (HT: NY!; IT: US!).

A. torreyi NAF **24:** 448. 1929. NEVADA: Empire City, 1865, *Torrey 93* (HT: NY!; IT: NY!).

A. williamsii Rydb. *in* Britton & Rydb., Bull. New York Bot. Gard. **2:** 175. 1901. **CANADA.** YUKON: Near Big Salmon, 22 Aug 1899, *Williams s.n.* (HT: NY! on 2 sheets; IT: US!).

Atelophragma BTBC **32:** 660. 1905 [1906]. T: *Astragalus aboriginorum* Richardson.

A. alpiniforme BTBC **55:** 129. 1928. OREGON: Wallowa Mts., Hurricane Creek, 25 Aug 1898, *Cusick 2103* (HT: GH!; IT: G, K, MIN, MO, NDG, ORE, P, US!, WS).

A. atratum NAF **24:** 372. 1929. **CANADA.** YUKON: Herschel Island, 22 Aug 1896, *Seale s.n.* (HT: DS!).

A. bracteatum BTBC **55:** 161. 1928. **MEXICO.** MÉXICO: Nevada de Toluca, 15 Oct 1903, *Rose & Painter 7913* (HT: NY!; IT: US!).

A. collieri BTBC **55:** 127. 1928. ALASKA: Eagle, 29 Jun 1902, *Collier 50* (HT: US!).

A. fernaldii BTBC **55:** 126. 1928. **CANADA.** LABRADOR: Blanc Sablon, 6 Aug 1910, *Fernald 3632* (HT: NY!; IT: CAN, GH 2 sheets).

A. harringtonii BTBC **55:** 126. 1928. ALASKA: Goodnews Bay, 14 Jul 1919, *Harrington 60* (HT: US!).

A. harshbergeri BTBC **55:** 160. 1928. **MEXICO.** Salazar, Sierra de las Cruces, 13 Aug 1896, *Harshberger 20* (HT: GH!; IT: US! no location).

A. herriotii BTBC **55:** 123. 1928. **CANADA.** MANITOBA: Qu'Appelle River, 28 Jun 1906, *Macoun & Herriot* (herbarium *#70479*) *s.n.* (HT: NY!; IT: CAS!).

A. hidalgense BTBC **55:** 159. 1928. **MEXICO.** HIDALGO: Sierra de la Mesa, 31 Jul & 1 Aug 1905, *Rose, Painter & Rose 9132* (HT: NY!; IT: GH, US!).

A. jaliscense BTBC **55:** 157. 1928. **MEXICO.** JALISCO: Sierra de Tequila, 5 Jul 1893, *Pringle 4430* (HT: NY!; IT: BM, G, K, NDG, OB, P, US!).

A. lineare BTBC **40:** 50. 1913. **CANADA.** YUKON: Foot of Lake Lebarge, 23 Jun 1899, *Tarleton 34b* (HT: NY!; IT: US!).

A. tioides BTBC **55:** 157. 1928. **MEXICO.** SAN LUIS POTOSÍ: Valley of San Luis Potosí, 1876, *Schaffner 822* (HT: GH!; IT: BM, CAS!, K, NY!, P, US!).

A. townsendii BTBC **55:** 163. 1928. **MEXICO.** CHIHUAHUA: Sierra Madre 10 miles southeast of Colonia García, 27 Jul 1899, *Townsend & Barber 95* (HT: NY!; IT: ARIZ, K, NDG, P, POM, RM!, US!).

A. wallowense BTBC **55:** 122. 1928. OREGON: Wallowa Mts., 28 Jul 1899, *Cusick 2267* (HT: GH!; IT: P, RM!).

A. zacatecanum BTBC **55:** 157. 1928. **MEXICO.** ZACATECAS: 26 Aug 1897, *Rose 2638* (HT: NY!; IT: GH, US!).

Batidophaca NAF **24:** 314. 1929. T: *Astragalus villosus* Michx.

B. humivagans NAF **24:** 316. 1929. ARIZONA: Mokiak Pass, 1877, *Palmer 108* (HT: GH!; IT: CAS!, NY!, US! no data).

B. sabinarum NAF **24:** 320. 1929. **UTAH. Iron Co.:** Cedar Canyon, 19 Jul 1920, *Garrett R2660* (HT: NY!).

B. stipulacea NAF **24:** 316. 1929. **NEW MEXICO. Grant Co.:** Mangas Springs, 12 Apr 1903, *Metcalfe 60* (HT: NY!; IT: MIN, NDG, NMC, US!).

Bembicidium Rydb. *in* Britton, Mem. Torrey Bot. Club **16:** 68. 1920. T: *B. cubense* Rydb.

B. cubense Rydb. *in* Britton, Mem. Torrey Bot. Club **16:** 68. 1920. **CUBA.** ORIENTE: Vicinity of Baracoa, 18 Feb–11 Mar 1910, *Shafer 4284* (HT: NY!).

Benthamantha brandegei NAF **24:** 246. 1924. **MEXICO.** LOWER CALIFORNIA: West slope of Cape Region Mts., Nov 1902, *Brandegee s.n.* (HT: US!; IT: UC).

B. chiapensis NAF **24:** 244. 1924. **MEXICO.** CHIAPAS: Near Jalisco, 12 Dec 1907, *Doyle 50* (HT: US!).

B. microphylla NAF **24:** 244. 1924. **MEXICO.** MICHOACÁN or GUERRERO: Río Coajaguillo, Feb 1899, *Langlassé 830* (HT: US!).

B. painteri NAF **24:** 246. 1924. **MEXICO.** JALISCO: Near Guadalajara, 30 Sep 1903, *Rose & Painter 7455* (HT: US!).

B. panamensis NAF **24:** 244. 1924. **PANAMA [COSTA RICA].** Hills around Alajuela, 11–12 Jan 1911, *Pittier 2340* (HT: US!).

B. robusta NAF **24:** 247. 1924. **MEXICO.** GUERRERO: Iguala Canyon, 31 Nov 1903, *Holway 40* (HT: US!).

B. trifoliata NAF **24:** 247. 1924. **MEXICO.** PUEBLA: Tehuacán, Jun 1912, *Purpus 5715* (HT: NY!; IT: F, GH, UC, US!).

B. wrightii NAF **24:** 246. 1924. ARIZONA. **Cochise Co.:** East slope of the Huachuca Mts., between the San Pedro and the Sonoita, *Wright 963* in part (HT: NY!; IT: UC, US! no data).

Brachyphragma Amer. J. Bot. **16:** 204. 1929. T: *Astragalus nudus* S. Wats.

Brongniartia alamosana NAF **24:** 192. 1923. **MEXICO.** SONORA: Alamos, 16–30 Sep 1890, *Palmer 658* (HT: US!).

B. angustata NAF **24:** 194. 1923. **MEXICO. [OAXACA]:** Between Mitla & Oaxaca City, 5–6 Sep 1906, *Rose & Rose 11295* (HT: NY).

B. argentea NAF **24:** 195. 1923. **MEXICO.** OAXACA: Miahuatlán, 17 May 1917, *Conzatti & Reko 3280* (HT: US!).

B. hirsuta NAF **24:** 194. 1923. **MEXICO.** QUERÉTARO:

Hacienda Ciervo, 20 Aug 1905, *Rose, Painter & Rose 9652* (HT: NY).

B. paniculata Rose ex Rydb., NAF **24:** 189. 1923. **MEXICO. Morelos:** Near Cuernavaca, 19 Nov 1895, *Pringle 7051* (HT: GH!).

B. pauciflora Rose ex Rydb., NAF **24:** 191. 1923. **MEXICO. Oaxaca:** Mouth of the Río Quintepec, 20 Jun 1919, *Conzatti & Cancino 2439* (HT: US!).

B. pringlei NAF **24:** 195. 1923. **MEXICO.** Tultenango, 13 Jul 1901, *Pringle 9431* (HT: NY!; IT: US!).

B. strigillosa NAF **24:** 192. 1923. **MEXICO. [Nayarit]:** Tepic, Acaponeta, 11 Apr 1910, *Rose, Standley & Russell 14395* (HT: US!; IT: NY!).

Cnemidophacos BTBC **32:** 663. 1905 [1906]. T: *Astragalus flavus* Nutt.

C. knowlesianus NAF **24:** 284. 1929. **Washington. Klickitat Co.:** 23 Apr & Jun 1886, *Suksdorf 482* (HT: NY!; IT: WS).

C. reventiformis NAF **24:** 284. 1929. **Washington.** Klickitat, 26 Apr 1882, *Howell s.n.* (HT: NY!; IT: DS!, ORE, WS).

C. urceolatus Fl. Rocky Mts. 502. 1917 [1918]. **Colorado:** Arboles, 3 Jun 1899, *Baker 424* (HT: NY!; IT: G, GH, K, MO, NDG, RM! 2 sheets, US).

Cracca arcuata NAF **24:** 166. 1923. **MEXICO. [Nayarit]:** Tres Marías Islands, 3–25 May 1897, *E. W. Nelson 4193* (HT: US!).

C. californica NAF **24:** 176. 1923. **MEXICO. Baja California:** Mesa Verde, 1899, *Brandegee s.n.* (HT: NY!).

C. calva NAF **24:** 161. 1923. **MEXICO. Jalisco:** Guadalajara, 11 Jun 1902, *Pringle 9773* (HT: NY!; IT: US!).

C. carpenteri NAF **24:** 172. 1923. **Florida:** Pensacola, Jun 1838, *Carpenter 44* (HT: NY!).

C. conzattii NAF **24:** 162. 1923. **MEXICO. Oaxaca:** Las Sedas, 19 May 1907, *Conzatti 1786* (HT: US!; IT: NY!).

C. curtissii Small ex Rydb., NAF **24:** 179. 1923. **Florida:** Cape Malabar, Aug–Sep 1879, *Curtiss 584* (HT: NY!; IT: NY! 2 sheets, US!).

C. foliolosa NAF **24:** 162. 1923. **MEXICO. Sinaloa:** Cerro Colorado, 3 Nov 1904, *Brandegee s.n.* (HT: US!).

C. hamata T. S. Brandegee ex Rydb., NAF **24:** 177. 1923. **MEXICO. Lower California:** San José del Cabo, 1 Oct 1890, *Brandegee s.n.* (HT: NY!; IT: DS!; CT: US!).

C. heydeana NAF **24:** 166. 1923. **GUATEMALA. Santa Rosa:** Cerro Redondo, Oct 1893, *Heyde & Lux 6111* (HT: NY!; IT: US!).

C. leucosericea NAF **24:** 163. 1923. False Washita, between Fort Cobb and Fort Arbuckle, 1868, *Palmer 114* (HT: NY!; IT: NY! 2 sheets).

C. mohrii NAF **24:** 163. 1923. **Florida:** Jun 1880, *Mohr s.n.* (HT: US!).

C. roseana NAF **24:** 164. 1923. **MEXICO. Querétaro:** Hacienda Ciervo, 20 Aug 1905, *Rose, Painter & Rose 9642* (HT: NY!; IT: US!).

C. rusbyi NAF **24:** 181. 1923. **MEXICO. Oaxaca:** Santa Catarina, 14 Jun 1910, *Rusby 69* (HT: NY!).

C. texana NAF **24:** 176. 1923. **Texas:** Hempstead, 1 Jun 1872, *E. Hall 119* (HT: NY!; IT: US!).

C. thurberi A. Gray ex Rydb., NAF **24:** 165. 1923. **MEXICO. Sonora:** Mububi, Jun 1851, *Thurber 410* (ST: NY! 2 sheets); Barbocomori, 18 Sep 1851, *Thurber 1009* (ST: NY!).

C. velutina NAF **24:** 171. 1923. **MEXICO. [Nayarit]:** Tepic, Zopelote, Feb 1895, *Lamb 575* (HT: NY!; IT: DS!, GH!, US!).

Ctenophyllum BTBC **32:** 663. 1905 [1906]. T: *Astragalus pectinatus* Dougl.

Cystium cornutum NAF **24:** 412. 1929. **Oregon:** John Day Valley, Muddy Station, 13 May 1885, *Howell s.n.* (HT: NY!; IT: F, ORE, P, UC).

C. griseolum NAF **24:** 411. 1929. **Nevada:** Candelaria, May 1882, *Shockley 220* (HT: GH!; IT: NDG).

C. heliophilum Fl. Rocky Mts. 491, 1063. 1917 [1918]. **Montana:** Lima, 1 Jul 1895, *Shear 3430* (HT: NY!; IT: RM! dated 1 Jul 1898).

C. kennedyi NAF 24: 407. 1929. **Nevada:** Carson Sink Region, Oct 1907, *Kennedy 1691* (HT: NY!; IT: DS! 3 sheets, RENO!, RM! 2 sheets, UC!, US!).

C. macrolobum NAF **24:** 408. 1929. **Nevada:** West Humboldt Mts., Jun 1868, *Watson 254* in part (HT: NY!; IT: GH!, US!, YU!).

C. merrillii NAF **24:** 410. 1929. **Wyoming:** Leucite Hills, 18 Jun 1901, *Merrill & Wilcox 680* (HT: NY!; IT: GH, RM!, US!).

C. ormsbyense NAF **24:** 412. 1929. **Nevada: Ormsby Co.:** Eagle Valley, 7 Jul 1902, *Baker 1275* (HT: NY!; IT: F, GH!, RSA!, UC!).

C. pardalotum NAF **24:** 415. 1929. **California:** Mohave Desert, May 1893, *Davidson s.n.* (HT: MIN—fide Ownbey in litt. 1984).

C. platyphyllidium NAF **24:** 410. 1929. **Oregon. Lilliam Co.:** Pine Creek, 7 Jun 1894, *Leiberg 171* (HT: NY!; IT: GH, ORE, US!).

C. sesquimetrale NAF **24:** 414. 1929. **Nevada:** Soda Springs, 11 Jun 1882, *Shockley 278* (HT: GH!; IT: NDG, UC!, US!; CT: NY!).

C. stramineum NAF **24:** 409. 1929. **Utah:** Southeast Utah, 1870, *Palmer s.n.* (HT: NY!; IT: US!).

C. tehachapiense NAF **24:** 414. 1929. **California. Kern Co.:** Tehachapi, 5 May 1905, *Heller 7833* (HT: NY!; IT: DS! 3 sheets, NY!, PH, US!, WIS).

C. vulpinum NAF **24:** 409. 1929. **Nevada:** Grass Valley, Jun 1868, *Watson 254* in part (HT: NY!; IT: GH!).

Daubentonia drummondii Amer. J. Bot. **10:** 498. 1923. **Texas:** *Drummond III/71* (HT: NY!; IT: US!).

Daubentoniopsis Amer. J. Bot. **10:** 497. 1923. T: *Aeschynomene longifolia* Cav.

Diholcos BTBC **32:** 664. 1905 [1906]. T: *Phaca bisulcata* Hook.

D. micranthus BTBC **34:** 420. 1907. **Colorado:** La Veta, 13 Jul 1896, *Shear 3569* (HT: NY!).

Diphysa microphylla NAF **24:** 213. 1924. **MEXICO. Tamaulipas:** Vicinity of Victoria, 1 Feb—9 Apr 1907, *Palmer 367* (HT: NY!; IT: US!).

D. puberulenta NAF **24:** 214. 1924. **MEXICO. [Nayarit]:** Acaponeta, Tepic, Feb 1895, *Lamb 520* (HT: NY!; IT: US! 2 sheets).

D. punctata NAF **24:** 211. 1924. **MEXICO. Morelos:** Cuernavaca, 9 Sep 1903, *Rose & Painter 6900* (HT: US!).

D. spinosa NAF **22:** 213. 1924. **MEXICO. Chiapas:** Canjob, 27 Mar 1904, *Goldman 796* (HT: US!).
D. villosa NAF **24:** 214. 1924. **MEXICO. Morelos:** Yantepec, 18 Jun 1901, *Pringle 9636* (HT: US!).
Geoprumnon Rydb. *in* Small, Fl. S.E. U.S. 615, 1332. 1903. T: *Astragalus crassicarpus* Nutt.
Gynophoraria NAF **24:** 280. 1929. T: *G. falcata* Rydb.
G. falcata NAF **24:** 280. 1929. **CANADA. Yukon:** Lake Kluane to Don Jek River, 11–27 Aug 1920, *Muller s.n.* (HT: PH—fide Schuyler in litt. 1984).
Hamosa asperula BTBC **54:** 335. 1927. **MEXICO. Hidalgo:** Between Pachuca and Real del Monte, 31 Aug 1903, *Rose & Painter 6679* (HT: US!).
H. atratiformis BTBC **34:** 48. 1907. **Utah:** Southern Utah, 1874, *Parry 47* (HT: NY!; IT: GH, ISC, K).
H. davisiana BTBC **54:** 328. 1927. **Texas:** Davis Mts., 28 Apr 1902, *Tracy & Evans 329* (HT: NY!; IT: NY! dated 15 Apr–3 May, US!).
H. emoryana BTBC **54:** 327. 1927. **Texas:** El Paso, 1852, *Wright 1359* (HT: NY!; IT: GH, MO).
H. gooddingii BTBC **54:** 20. 1927. **Arizona:** Huachuca Mts., May 1912, *Goodding 1299* (HT: NY!).
H. hidalgensis BTBC **54:** 333. 1927. **MEXICO. Hidalgo:** Dublán, 19 Sep 1902, *Pringle 9720* (HT: NY!; IT: GH, K, US!).
H. imperfecta BTBC **54:** 329. 1927. **Nevada. Nye Co.:** Rhyolite, 17 May 1909, *Heller 9637* (HT: NY!; IT: RENO!).
H. manca BTBC **54:** 17. 1927. **Nevada. Elko Co.:** Clover Mts., 24 Jul 1908, *Heller 9163* (HT: NY!; IT: US!).
H. minthorniae BTBC **54:** 15. 1927. **Nevada. Lincoln Co.:** Vicinity of Pioche, 24 Apr 1909, *Minthorn 77* (HT: NY!, IT: UC!).
H. monophyllidia NAF **24:** 421. 1929. **Nevada. Nye Co.:** Currant, May 1916, *Bentley s.n.* (HT: NY!; IT: DS!, MO, RM! dated 15 May, RSA!).
H. montereyensis BTBC **54:** 326. 1927. **MEXICO. Nuevo León:** Monterey, Feb 1880, *Palmer 237* (HT: GH!).
H. paysonii BTBC **54:** 22. 1927. **Wyoming. Sublette Co.:** Horse Creek, seven miles west of Merna, 18 Jul 1922, *Payson & Payson 2748* (HT: NY!; IT: GH, MO, RM!, US!).
Hedysarum lancifolium Mem. New York Bot. Gard. **1:** 256. 1900. **Montana:** Headwater of Jocko River, 11 Jul 1883, *Canby 93* (HT: NY!).
H. utahense BTBC **34:** 424. 1907. **Utah:** Vicinity of Salt Lake City, 26 May 1883, *Leonard 55* (HT: NY!).
Hesperastragalus milesianus BTBC **53:** 169. 1926. **California:** San Luis Obispo, Apr 1886, *Miles s.n.* (HT: NY!).
H. obispensis BTBC **53:** 167. 1926. **California:** San Luis Obispo, 6 May 1882, *Jones 3229* in part (HT: NY!; IT: POM).
Hesperonix NAF **24:** 438. 1929. T: *Astragalus bolanderi* A. Gray.
Hesperothamnus brachycalyx NAF **24:** 237. 1924. **MEXICO. Oaxaca:** Six miles above Dominguillo, 30 Oct 1894, *E. W. Nelson 1826a* (HT: US!).
Hoita NAF **24:** 7. 1919. T: *Psoralea macrostachya* DC.
H. hallii NAF **24:** 10. 1919. **California. Butte Co.:** Humboldt Road, 31 Jul 1914, *H. M. Hall 9825* (HT: UC!; IT: CAS!; CT: NY!).
H. longiloba NAF **24:** 10. 1919. **California:** 1876, *Parry & Lemmon 84* (HT: NY!; IT: CAS!).
H. villosa NAF **24:** 10. 1919. **California. Sonoma Co.:** Burke's Sanitarium, 24 Jun 1902, *Heller 5761* (HT: NY!; IT: DS!, US).
Holcophacos Rydb. *in* Small, Fl. S.E. U.S. 618, 1332. 1903. T: *Astragalus distortus* Torrey & Gray.
Homalobus amblyodon NAF **24:** 266. 1929. **Alaska:** Stepovak Bay, 8–19 Jul 1899, *Palache s.n.* (HT: US!; CT: NY!).
H. camporum BTBC **32:** 666. 1905 [1906]. **Wyoming. Sweetwater Co.:** Bush Ranch, summit of Steamboat Mt., 10 Jun 1900, *Nelson 7085* (HT: NY!; IT: RM! 2 sheets).
H. canovirens Fl. Rocky Mts. ed. 2, 1126. 1922 [1923]. **Utah:** La Sal Mts., near Little Springs, 5–6 Jul 1911, *Rydberg & Garrett 8536* (HT: NY!).
H. clementis BTBC **31:** 563. 1904. **Colorado:** Sangre de Cristo, 1896, *Clements 354* (HT: NY!).
H. decurrens BTBC **31:** 563. 1904. **Colorado:** Estes Park, Aug 1895, *Osterhout s.n.* (HT: NY!; IT: RM! dated 10 Aug and numbered *730*).
H. hitchcockii NAF **24:** 271. 1929. **Nevada:** Duck Creek near Ely, 17 Aug 1913, *Hitchcock 1344* (HT: US!).
H. humilis BTBC **34:** 417. 1907. **Utah:** Mountain north of Bullion Creek, near Marysvale, 23 Jul 1905, *Rydberg & Carlton 7147* (HT: NY!; IT: GH, RM!, US!).
H. hylophilus Mem. New York Bot. Gard. **1:** 247. 1900. **Montana:** Bridger Mts., 17 Jun 1897, *Rydberg & Bessey 4490* (HT: NY!; IT: CAN, GH, K, NDG, NEB, PH, RM!, US!).
H. laurentii BTBC **51:** 15. 1924. **Oregon:** East of Heffner, 17 Jul 1917, *Lawrence 744* (HT: NY!; IT: RSA, US!).
H. macgregorii BTBC **50:** 270. 1923. **California. Ventura Co.:** Near Frazier Borax Mine, Mt. Pinos, 12–14 Jun 1908, *Abrams & McGregor 219* (HT: NY!; IT: DS!, G, US!).
H. microcarpus BTBC **34:** 417. 1907. **Colorado:** East slope of Rabbit Ear Range, 20 Jul 1894, *collector unknown* (from Herb. State Agric. College, Colo.) *s.n.* (HT: NY!).
H. oblongifolius BTBC **34:** 50. 1907. **Colorado:** Cerro Summit, 12 Jul 1901, *Baker 409* (HT: NY!; IT: RM! 2 sheets, US!, WS).
H. paucijugus BTBC **34:** 418. 1907. **Utah:** Big Cottonwood Canyon near the summit of the divide between Lake Solitude and Twin Lakes, 9 Aug 1905, *Garrett 1580* (HT: NY!; IT: US!).
H. plummerae BTBC **51:** 16. 1924. **California. Sierra Co.:** 1875, *Lemmon 622* (HT: NY!; IT: GH).
H. proximus BTBC **32:** 667. 1905 [1906]. **Colorado:** Arboles, Jun 1899, *Baker 421* (HT: NY!; IT: G, GH, K, MO, POM, RM! 2 sheets, US!).
H. retusus BTBC **50:** 186. 1923. **CANADA. Yukon:** Fort Selkirk, 20 Jul 1899, *Tarleton 128a* (ST: NY!, US!); *128b* (ST: NY!).
H. salidae BTBC **32:** 667. 1905 [1906]. **Colorado:** Salida, 3 Aug 1896, *Shear 3468* (HT: NY!).
H. standleyi BTBC **50:** 182. 1923. **New Mexico:** Pon-

chuelo Creek, 4 Jul 1908, *Standley 4181* (HT: NY!; IT: US!).

H. stipitatus BTBC **34:** 419. 1907. **North Dakota:** Upper Missouri, 1839, *Geyer s.n.* (HT: NY!).

H. strigulosus BTBC **34:** 420. 1907. **Nevada:** East Humboldt Mts., Jul 1868, *Watson 283* (HT: NY!; IT: GH!, YU!).

H. subglaber BTBC **51:** 17. 1924. **Oregon:** John Day River, 10 May 1882, *Howell s.n.* (HT: NY!; IT: DS!, US!).

H. uniflorus BTBC **34:** 49. 1907. **Wyoming:** Evanston, 28 May 1897, *Nelson 2971* (HT: NY!; IT: RM!, UTC!).

H. wolfii BTBC **31:** 562. 1904. **Colorado:** South Park, Jul 1873, *Wolf 243* (HT: NY!; IT: GH!).

Indigofera argentata NAF **24:** 143. 1923. **MEXICO. Coahuila:** Sabinos, 21 May 1902, *E. W. Nelson 6833* (HT: US!).

I. discolor NAF **24:** 147. 1923. **MEXICO. Sinaloa:** Lodiego, 9–15 Oct 1891, *Palmer 1634* (HT: NY!; IT: US! 2 sheets).

I. laevis NAF **24:** 144. 1923. **MEXICO. Sonora:** Guaymas, 1887, *Palmer 296* (HT: GH!; IT: DS!, NY!, US!).

I. lancifolia NAF **24:** 147. 1923. **MEXICO. [Guerrero]:** Acapulco, Oct 1894–Mar 1895, *Palmer 251* (HT: US!; IT: NY!).

I. langlassei NAF **24:** 147. 1923. **MEXICO. Guerrero or Michoacán:** Sierra Madre, 9 Nov 1898, *Langlassé 609* (HT: US!).

I. nana NAF **24:** 143. 1923. **MEXICO. Michoacán:** Sunguato, 24 Jun 1910, *Arsène s.n.* (HT: NY!).

I. nelsonii NAF **24:** 150. 1923. **MEXICO. Lower California:** La Laguna and El Taraiso, 29 Jan 1906, *E. W. Nelson & Goldman 7472* (HT: US!).

I. panamensis NAF **24:** 147. 1923. **PANAMA.** Penonomé, 23 Feb–22 Mar 1908, *Williams 123* (HT: NY!).

I. pueblensis NAF **24:** 150. 1923. **MEXICO. Puebla:** Near San Luis Tultitlanapa, Jul 1907, *Purpus 2672* (HT: NY!; IT: US!).

I. rosei NAF **24:** 145. 1923. **MEXICO. Zacatecas:** San Juan Capistrano, 22 Aug 1897, *Rose 2480* (HT: NY!; IT: US!).

I. sphenoides NAF **24:** 142. 1923. **MEXICO. San Luis Potosí:** San Dieguito, 13–16 Jun 1904, *Palmer 95* (HT: GH!; IT: US!).

Jonesiella BTBC **32:** 661. 1905 [1906]. T: *Astragalus asclepiadoides* M. E. Jones.

J. ellisiae NAF **24:** 403. 1929. **New Mexico:** Sandia Mts., 30 Apr 1914, *Ellis 421* (HT: US!; IT: MO; CT: NY!).

J. mearnsii NAF **24:** 403. 1929. **Arizona:** Fort Verde, May 1887, *Mearns 163* (HT: NY!).

J. recedens E. L. Greene ex Rydb., NAF **24:** 403. 1929. **Colorado:** Arboles, Jun 1899, *Baker 422* (HT: NY!; IT: GH, K, MO, RM! 2 sheets).

Kentrophyta minima BTBC **34:** 420. 1907. **Wyoming:** Yellowstone National Park, Aug 1884, *Tweedy 83* (HT: NY!).

Kuhnistera adenopoda NAF **24:** 136. 1920. **Florida. [Hillsborough Co.]:** Tampa, Oct 1877, *Garber s.n.* (HT: GH!; IT: NY!, US!).

K. candida var. *diffusa* Fl. Nebraska **21:** 59. 1895. **Nebraska. Deuel Co.:** 25 Jun 1891, *Rydberg 58* (HT: NEB—fide Bolick in litt. 1985; IT: NY!).

Lathyrus brachycalyx BTBC **34:** 425. 1907. **Utah:** City Creek Canyon, 9 Jun 1883, *Leonard 101* (HT: NY!).

L. laetivirens E. L. Greene ex Rydb., Fl. Colorado 217. 1906. **Colorado:** Cerro Summit, 17 Jun 1901, *Baker 149* (LT: NY!—C. L. Hitchcock, Univ. Wash. Publ. Biol. **15:** 39. 1952; ILT: RM! 2 sheets).

L. leucanthus BTBC **28:** 37. 1901. **Colorado:** Ojo, 26 May 1900, *Rydberg & Vreeland 6020* (HT: NY!; IT: CS, NEB, RM!).

L. ornatus var. *flavescens* Fl. Nebraska **21:** 64. 1895. **Nebraska. Dodge Co.:** Fremont, May 1889, *Rydberg s.n.* (HT: NEB—fide Bolick in litt. 1985).

L. ornatus var. *incanus* Smith & Rydb. *in* Rydb., Fl. Nebraska **21:** 65. 1895. **Nebraska. Sheridan Co.:** 11 Jul 1892, *Smith & Pound 49* (HT: NEB—designated by J. G. Smith in herb., fide Bolick in litt. 1985; IT: NEB).

Lonchophaca NAF **24:** 312. 1929. T: *Phaca macrocarpa* A. Gray.

Lotus longebracteatus BTBC **30:** 254. 1903. **Utah:** South Utah, 1877, *Palmer 94* (HT: US!).

Lupinus adscendens BTBC **30:** 256. 1903. **Wyoming:** Headwaters of the Tongue River, Jul 1898, *Tweedy 129* (HT: NY!; IT: YU! dated 13 Jul).

L. argentinus BTBC **30:** 257. 1903. **Nevada** [Utah in publication]: Near Reno, 19 Jun 1900, *Stokes s.n.* (HT: NY!; IT: US!).

L. candicans BTBC **28:** 35. 1901. **Montana:** 1900, *Wilcox 451* (HT: US; IT: NY!).

L. comatus BTBC **30:** 257. 1903. **Colorado:** Lake City, 10 Jun 1878, *Pease s.n.* (HT: NY!).

L. cyaneus BTBC **28:** 35. 1901. **Montana:** 4 Jul 1900, *Wilcox 446* (HT: US!; IT: NY!).

L. decumbens var. *argentatus* Fl. Colorado 197–199. 1906. **Colorado. Larimer Co.:** Jul 1896, *Baker s.n.* (LT: NY!—Barneby, Brittonia **41:** 154. 1989).

L. depressus BTBC **30:** 255. 1903. **Idaho:** Coeur d'Alene Mts., divide between St. Joseph and Clearwater Rivers, 9 Jul 1895, *Leiberg 1201* (HT: US!; IT: NY!).

L. evermannii BTBC **30:** 255. 1903. **Idaho:** Near Sawtooth, 9–10 Jul 1896, *Evermann 533* (HT: US!; IT: DS!, NY!).

L. flavescens BTBC **29:** 245. 1902. **Idaho or Montana:** Medicine Clay Prairies, 10 Jun, *Wyeth s.n.* (HT: NY!).

L. flavicaulis BTBC **34:** 43. 1907. **Wyoming:** Snake River, 19 Aug 1894, *Nelson 1098* (HT: NY!; IT: RM! labeled Gros Ventre River and dated 16 Aug).

L. jonesii BTBC **30:** 256. 1903. **Utah:** Silver Reef, 3 May 1894, *Jones 5143* (HT: US!; IT: BRY, NY!, RM!).

L. laxispicatus BTBC **34:** 42. 1907. **Idaho. Kootenai Co.:** Jul 1887, *Sandberg s.n.* (HT: NY!).

L. laxus BTBC **30:** 258. 1903. **Montana:** Forks of the Madison, 26 Jul 1897, *Rydberg & Bessey 4442* (HT: NY!).

L. leucanthus BTBC **30:** 259. 1903. **Utah:** Springdale, 16 May 1894, *Jones 5249e* (HT: US!; IT: BRY, DS!, NY!, POM, RM!; CT: NY!).

L. lucidulus BTBC **29:** 245. 1902. **Wyoming:** Spread Creek, Aug 1897, *Tweedy 271* (HT: NY!; IT: YU! dated 15 Jul–15 Aug).

L. lupinus BTBC **40:** 44. 1913. **Utah:** Elk Mts., western

Bear's Ear, 2 Aug 1911, *Rydberg & Garrett 9363* (HT: NY!).

L. macounii BTBC **34:** 42. 1907. **CANADA. Saskatchewan:** Assiniboia, Cypress Hills, 30 Jun 1894, *Macoun 4070* (HT: NY!; IT: CAN).

L. macrostachys BTBC **34:** 44. 1907. **Montana:** Jocko Creek, 16 Jun 1901, *MacDougal 253* (HT: NY!; IT: US!).

L. maculatus BTBC **30:** 257. 1903. **Utah:** Wasatch Mts., Pleasant Valley Junction, Aug 1883, *Jones s.n.* (HT: NY!; IT: CAS!, NY! 2 sheets, POM, US! 2 sheets, UTC!).

L. marianus BTBC **34:** 41. 1907. **Utah:** Along Bullion Creek above Marysvale, 21 Jul 1905, *Rydberg & Carlton 7024* (HT: NY!; IT: CAN, RM!, US!).

L. monticola Mem. New York Bot. Gard. **1:** 232. 1900. **Montana:** Indian Creek, 22 Jul 1897, *Rydberg & Bessey 4447* (HT: NY!; IT: CAN, MONT, NEB 2 sheets, NY!, RM!, US!).

L. pseudoparviflorus Mem. New York Bot. Gard. **1:** 232. 1900. **Montana:** Bridger Mts., 17 Jun 1897, *Rydberg & Bessey 4441* (HT: NY!; IT: CAN, MONT, NEB 2 sheets, RM!, US!).

L. pulcherrimus BTBC **30:** 258. 1903. **Wyoming:** Battle, Continental Divide, 20 Jun 1901, *Tweedy 4215* (HT: NY!; IT: US!, YU!).

L. roseolus BTBC **34:** 44. 1907. **Wyoming:** Continental Divide, Buffalo Fork, Aug 1897, *Tweedy 270* (HT: NY!; IT: YU! dated 15 Jul–15 Aug).

L. rubens BTBC **34:** 45. 1907. **Utah:** Southern Utah, 1874, *Parry 41* (HT: NY!; IT: ISC, NY!).

L. scaposus BTBC **34:** 45. 1907. **Colorado. Garfield Co.:** Glenwood Springs, 17 Jun 1899, *Osterhout s.n.* (HT: NY!; IT: RM!).

L. scheuberae BTBC **29:** 244. 1902. **Montana. Garnet Co.:** 1901, *Sheuber 135* (HT: NY!; IT: MONT dated 20 Jul 1901).

L. spathulatus BTBC **29:** 244. 1902. **Utah:** Wasatch Mts., Jun 1869, *Watson 225* (HT: NY!).

L. stenophyllus BTBC **34:** 42. 1907. Oregon Plains, *Nuttall s.n.* (HT: not found at NY).

L. subulatus BTBC **34:** 43. 1907. **Montana:** Columbia Falls, 3 Jun 1897, *Williams s.n.* (HT: NY!; IT: RM!).

Microphacos BTBC **32:** 663. 1905 [1906]. T: *Astragalus gracilis* Nutt.

Parosela abietifolia Rose ex Rydb., NAF **24:** 92. 1920. **MEXICO. Jalisco:** Etzatlán, 27 Oct 1903, *Pringle 8774* (HT: US!; IT: CAS!, F, GH, NY!, UC).

P. alamosana Rose ex Rydb., NAF **24:** 63. 1919. **MEXICO. Sonora:** Alamos, 16–30 Sep 1890, *Palmer 711* (HT: US!; IT: F, GH, NY!).

P. atrocyanea NAF **24:** 111. 1920. **MEXICO. Puebla:** San Luis Tultitlanapa, Jul 1908 *Purpus 2651* (HT: NY!; IT: F, GH, UC, US!).

P. attenuata NAF **24:** 82. 1920. **MEXICO. Oaxaca:** Sierra de San Felipe, 26 Oct 1894, *C. L. Smith 547* (HT: NY!; IT: F, NY!, US!).

P. barberi Rose ex Rydb., NAF **24:** 74. 1920. **MEXICO. Chihuahua:** Sierra Madres near Colonia García, 22 Aug 1899, *Townsend & Barber 275* (HT: US!; IT: NY!).

P. bigelovii NAF **24:** 75. 1920. **NEW MEXICO.** Santa Rita del Cobre, *Bigelow 266* in part (HT: US!).

P. botterii NAF **24:** 110. 1920. **MEXICO. [Veracruz]:** Orizaba, *Botteri 973* (HT: GH!; CT: NY!).

P. capitulata NAF **24:** 113. 1920. **MEXICO. Oaxaca:** 8 Dec 1901, *Conzatti & Gonzales 1262* (HT: GH!; CT: NY!).

P. carnescens Fl. Rocky Mts. 483, 1063. 1917 [1918]. **[New Mexico]:** Santa Rita del Cobra, 5 Nov 1851, *Bigelow s.n.* (LT: NY!—Barneby, Mem. New York Bot. Gard. **27:** 566. 1977).

P. caudata NAF **24:** 82. 1920. **MEXICO.** 1853, *Müller 844* (HT: NY!).

P. caudata [biz] NAF **24:** 112. 1920. **MEXICO. San Luis Potosí:** Alvarez, 5–10 Sep 1902, *Palmer 110* (HT: US!; IT: NY!).

P. citrina NAF **24:** 81. 1920. **MEXICO. Jalisco:** Tuxpan, 27 Oct 1904, *Pringle 8860* (HT: NY!; IT: CAS!, F, GH, UC, US!).

P. conzattii NAF **24:** 89. 1920. **MEXICO. Oaxaca:** Cerro San Antonio, 15 Dec 1901, *Conzatti s.n.* (HT: US!; CT: NY!).

P. costaricana NAF **24:** 77. 1920. **COSTA RICA.** Haciendas de Zent et de Guacimo, United Fruit Company *#21* (HT: US!).

P. dispansa NAF **24:** 59. 1919. **MEXICO. Jalisco:** Talpa, 7 Mar 1897, *E. W. Nelson 4034* (HT: US!; IT: NY!).

P. diversicolor NAF **24:** 109. 1920. **MEXICO. Oaxaca:** Sierra de Clavellinas, 16–19 Oct 1894, *C. L. Smith 549* (HT: US!; IT: NY! 2 sheets one dated 19 Oct).

P. exserta NAF **24:** 73. 1920. **MEXICO. Colima:** 27–28 Feb 1891, *Palmer 1312* (HT: US!).

P. fissa NAF **24:** 74. 1920. **MEXICO. [Nayarit]:** Tepic, Sierra Madre near Santa Teresa, 11 Aug 1897, *Rose 2181* (HT: US!; IT: GH, NY!).

P. floridana NAF **24:** 114. 1920. **Florida:** Miami, Nov 1878, *Garber s.n.* (HT: NY!; IT: NY! 5 sheets).

P. fulvosericea NAF **24:** 89. 1920. **MEXICO. San Luis Potosí:** 1878, *Parry & Palmer 148* in part (HT: US!).

P. fuscescens NAF **24:** 109. 1920. **MEXICO. Coahuila:** Feb–Oct 1880, *Palmer 209* (HT: US!; IT: NY!).

P. gigantea Rose ex Rydb., NAF **24:** 110. 1920. **MEXICO. Michoacán:** Tarascon, 11 Oct 1904, *Pringle 8848* (HT: US!; IT: CAS!, NY!).

P. glabrescens NAF **24:** 106. 1920. **MEXICO. San Luis Potosí:** 1878, *Parry & Palmer 150* (HT: US!; IT: GH, NY!).

P. involuta NAF **24:** 107. 1920. **MEXICO. Jalisco:** Río Blanco, Jun–Oct 1886, *Palmer 26* (HT: NY!; IT: GH, NY!, US!).

P. lagopina NAF **24:** 72. 1920. **MEXICO. Durango:** Chacala, 7 Mar 1899, *Goldman 357* (HT: US!; IT: GH, NY!).

P. laxa NAF **24:** 85. 1920. **Texas:** Devil's River, 16 Oct 1913, *Rose & Fitch 17957* (HT: NY!; IT: US!).

P. leucantha NAF **24:** 103. 1920. **MEXICO. [Nayarit]:** Tepic, 1891–1892, *Palmer p* in part (HT: NY!; IT: GH, US! dated 5 Jan–6 Feb 1892).

P. leucosericea NAF **24:** 104. 1920. **MEXICO. Oaxaca:** Hills near Tule, 20 May 1906, *Pringle 13866* (HT: US!).

P. lloydii NAF **24:** 84. 1920. **MEXICO. Zacatecas:** Cedros, 29 Apr 1908, *Lloyd 60* (HT: US!; IT: UC).

P. longifolia Rose ex Rydb., NAF **24:** 105. 1920. **MEX-**

ICO. **Jalisco:** Road from Altenquilla to Jacala, 5 Mar 1897, *E. W. Nelson 4016* (HT: US!; IT: NY!).

P. lucida Rose ex Rydb., NAF **24:** 74. 1920. **MEXICO. Durango:** Cajetes, 1896, *Palmer 955* (HT: US!).

P. megalostachys Rose ex Rydb., NAF **24:** 102. 1920. **MEXICO. Lower California:** San Estebán, 17 Apr 1889, *Brandegee s.n.* (HT: US!; IT: DS! no date, NY!, UC).

P. minutifolia NAF **24:** 87. 1920. **MEXICO. Hidalgo:** Pachuca, May 1905, *Purpus 1145* (HT: NY!).

P. mollissima NAF **24:** 64. 1919. **Nevada:** Las Vegas Wash, 22 Apr 1905, *Goodding 2237* (HT: NY!; IT: GH, RM!, UC).

P. occidentalis NAF **24:** 115. 1920. **MEXICO. Sinaloa:** Culiacán, 25 Oct–18 Nov 1891, *Palmer 1783* (HT: NY!; IT: UC, US! labeled Lodiego and dated 9–15 Oct).

P. oculata NAF **24:** 60. 1919. **MEXICO. Lower California:** Cerralvo, 19 Apr 1911, *Rose 16884* (HT: US!).

P. pilifera NAF **24:** 75. 1920. **MEXICO. Hidalgo:** Pachuca, Sep 1905, *Purpus 1741* in part (HT: NY!; IT: US!).

P. pilosa NAF **24:** 64. 1919. **MEXICO. Lower California:** Los Angeles Bay, 1887, *Palmer 550* (HT: NY!; IT: GH, US!).

P. pilosissima NAF **24:** 91. 1920. **MEXICO.** Durango and vicinity, Apr–Nov 1896, *Palmer 808* (HT: GH!; IT: F, NY!, UC, US!).

P. platyphylla NAF **24:** 115. 1920. **MEXICO. Sinaloa:** Fuerte, 26 Mar 1910, *Rose, Standley & Russell 13527* (HT: US!; IT: NY!).

P. polycephala NAF **24:** 90. 1920. **MEXICO. San Luis Potosí:** 1878, *Parry & Palmer 160* (HT: US; IT: K, NY!, P).

P. roseiflora NAF **24:** 105. 1920. **MEXICO.** Sierra Madre, *Seeman s.n.* (HT: GH!; IT: K).

P. roseola NAF **24:** 81. 1920. **GUATEMALA. Santa Rosa:** Volcán Jumaytepeque, Nov 1892, *Heyde & Lux* [*D. Smith* distribution number] *4166* (HT: NY!; IT: GH, US!).

P. sanctae-crucis NAF **24:** 103. 1920. **MEXICO. Sonora:** Santa Cruz, 1851, *Wright 986* (HT: GH; IT: NY!, UC, US!).

P. sericocalyx NAF **24:** 63. 1919. **MEXICO.** Sierra Madre, *Seemann s.n.* (HT: GH!; IT: BM numbered *2182,* K; CT: NY!).

P. smithii NAF **24:** 106. 1920. **MEXICO. Oaxaca:** Sierra de San Felipe, 19 Nov 1894, *C. L. Smith 343* (HT: US!; IT: F, NY!, UC).

P. subvillosa NAF **24:** 93. 1920. **MEXICO. Chihuahua:** Near Paso del Norte, 23 Sep 1886, *Pringle 720* in part (HT: NY!; IT: NY!, US!).

P. tehuacana NAF **24:** 113. 1920. **MEXICO. [Puebla]:** Tehuacán, 7 Nov 1903, *Holway 53* (HT: US!).

P. townsendii NAF **24:** 76. 1920. **MEXICO. Chihuahua:** Sierra Madre near Colonia García, 4 Sep 1899, *Townsend & Barber 301* (HT: US!; IT: F, K, NY!, Z).

P. tsugoides NAF **24:** 102. 1920. **MEXICO. Oaxaca:** Valley of Oaxaca, 3 Oct 1894, *E. W. Nelson 1540* (HT: US!; IT: NY!).

P. tuberculina NAF **24:** 89. 1920. **MEXICO. Puebla:** San Luis Tultitlanapa, Jul 1908, *Purpus 3215* (HT: NY!; IT: F, GH, UC, US!).

P. variegata NAF **24:** 55. 1919. **MEXICO. Lower California:** San José del Cabo, 24 Oct 1902, *Brandegee s.n.* (HT: US!).

P. wardii NAF **24:** 112. 1920. **MEXICO. San Luis Potosí:** Sep 1891, *Ward s.n.* (HT: US!).

Pediomelum NAF **24:** 17. 1919. T: *Psoralea esculenta* Pursh.

P. brachypus NAF **24:** 24. 1919. **MEXICO. Oaxaca:** Los Naranjos, Aug 1908, *Purpus 3049* (HT: NY!; IT: DS!).

P. caudatum NAF **24:** 19. 1919. **Texas. Dallas Co.:** May, *Reverchon 563* (HT: NY!; IT: NY!).

P. humile NAF **24:** 24. 1919. **MEXICO. Coahuila:** Near Piedras Negras, 19 Apr 1910, *Pringle 9205* (HT: US).

P. retrorsum NAF **24:** 22. 1919. **Arizona:** Peach Springs, Jun 1884, *Lemmon & Lemmon s.n.* (HT: NY!).

P. sonorae NAF **24:** 23. 1919. **MEXICO. Sonora:** Alamos, 17 Mar 1910, *Rose, Standley & Russell 13025* (HT: NY!).

P. trinervatum NAF **24:** 23. 1919. **MEXICO. Chihuahua:** 5–10 Jun 1908, *Palmer 356* (HT: NY!).

Petalostemon confusus NAF **24:** 129. 1920. **MEXICO. Chihuahua:** Base of Sierra Madre, 20 Sep 1887, *Pringle 1215* (HT: US!; IT: CAS, F, K, NY!, UC).

P. lagopus NAF **24:** 134. 1920. **Nevada. Washoe Co.:** Truckee Pass, 14 Jun 1906, *Kennedy 1316* (HT: NY!; IT: RENO! 3 sheets, UC, US!).

P. molle Mem. New York Bot. Gard. **1:** 238. 1900. **Montana:** Camerons Ranch south of Snowy Mts., 17 Aug 1882, *Canby s.n.* (LT: NY!—Rydberg, NAF 24: 131. 1920; ILT: GH, US!).

P. obreniformis NAF **24:** 130. 1920. **MEXICO. Chihuahua:** Base of Mt. Mohinora, 8 miles from Guadalupe y Calvo, 23–31 Aug 1898, *E. W. Nelson s.n.* (HT: US!; IT: K, NY!).

P. pilulosus NAF **24:** 128. 1920. **Arizona:** 1891, *Nealley 237* (HT: NY!).

P. rothrockii NAF **24:** 134. 1920. **Arizona:** 1874, *Rothrock s.n.* (HT: NY!; IT: NY! numbered *81*).

P. sonorae NAF **24:** 125. 1920. **MEXICO. Sonora:** Río de Santa Cruz, 25 Jun 1855, *Schott 241* in part (HT: NY!; IT: F, NY!).

P. standleyanus NAF **24:** 131. 1920. **New Mexico:** Raton Mts., 18–19 Aug 1903, *Griffiths 5464* (HT: US!).

P. truncatus NAF **24:** 124. 1920. **MEXICO. Durango:** Tepehuanes, 4–25 Jun 1906, *Palmer 284* (HT: NY!; IT: US!).

Phaca arenicola NAF **24:** 356. 1929. **California:** Mecca, Colorado Desert, 28 Feb 1913, *Parish 8467* (HT: NY! on 2 sheets; IT: DS!).

P. argentata NAF **24:** 340. 1929. **Nevada:** West Humboldt Mts., Star Peak, 31 Jul 1912, *Heller 10633* (HT: NY!; IT: CAS!, RENO!).

P. davidsonii NAF **24:** 362. 1929. **California:** Lancaster, 12 May 1893, *Davidson 4* (HT: GH!; IT: DS, NY!, US! not numbered).

P. deanei NAF **24:** 355. 1929. **California. San Diego**

Co.: Sweetwater Valley, 13 May 1883, *Deane s.n.* (HT: GH!).

P. encenadae NAF **24:** 336. 1929. **MEXICO. [BAJA CALIFORNIA]:** Encenada [Ensenada], 12 Apr 1882, *Jones 3675* [*3672* in publication] (HT: NY!; IT: CAS!, DS!, NY!, US!).

P. jucunda Jepson & Rydb. *in* Rydb., NAF **24:** 339. 1929. **CALIFORNIA:** Panamint Mts., Telescope Peak, 13 May 1917, *Jepson 7005* (HT: JEPS!).

P. lanosissima NAF **24:** 357. 1929. **CALIFORNIA. Los Angeles Co.:** La Bolsa, Oct 1882, *Parish & Parish 1117* (HT: NY!; IT: US! as Bullario Ranch?).

P. lenophylla NAF **24:** 341. 1929. **CALIFORNIA. Placer Co.:** Tinker's Knob, 24 Jul 1892, *Sonne s.n.* (HT: NY!).

P. lerdoensis NAF **24:** 356. 1929. **MEXICO. SONORA:** Colonia Lerdo, Feb 1904, *MacDougal s.n.* (HT: NY!).

P. megalophysa NAF **24:** 344. 1929. **CALIFORNIA:** San Antonio Mts., Swarthout Canyon, 3–6 Jun 1900, *H. M. Hall 1531* (HT: GH!).

P. nivea NAF **24:** 348. 1929. **MEXICO. LOWER CALIFORNIA:** 15 miles north of San Felipe Bay, Feb 1904, *MacDougal s.n.* (HT: NY!).

P. pardalina NAF **24:** 352. 1929. **UTAH.** Cedar Mt., Green River, 20 May 1915, *Jones s.n.* (HT: NY!; IT: DS, POM, UTC!).

P. perstricta NAF **24:** 344. 1929. **CALIFORNIA. San Diego Co.:** Hill Valley between Campo and Jacumba, 28 May 1903, *Abrams 3636* (HT: NY!; IT: DS! 2 sheets, PH, POM, US!).

P. potosina NAF **24:** 337. 1929. **MEXICO.** San Luis Potosí to Tampico, Dec 1878–Feb 1879, *Palmer 1051* (HT: PH—fide Schuyler in litt. 1984; IT: NY!).

P. prebblei NAF **24:** 361. 1929. **CANADA. ALBERTA:** Athabasca River, 14 Aug 1903, *Prebble & Cary 123* (HT: US!).

P. pseudocarpa NAF **24:** 343. 1929. **CALIFORNIA. San Diego Co.:** Cuiamaca Mts., 1875, *Palmer 68* (HT: NY!).

P. siskiyouensis NAF **24:** 340. 1929. **CALIFORNIA. Siskiyou Co.:** Mt. Eddy, 30 Aug 1912, *Eastwood 2006* (HT: NY!; IT: CAS!, US!).

P. topoensis NAF **24:** 343. 1929. **MEXICO. LOWER CALIFORNIA:** Topo, Oct 1882, *Orcutt 621* (HT: GH!).

P. tracyi NAF **24:** 351. 1929. **TEXAS:** Toyah Creek, 21 Apr 1902, *Tracy & Earle 78* (HT: NY!; IT: NDG, MO, TEX, US!).

P. tularensis NAF **24:** 358. 1929. **CALIFORNIA:** Tulare Lake, *Lemmon s.n.* (HT: GH!; CT: NY!).

P. unde NAF **24:** 346. 1929. **WASHINGTON:** Walla Walla, Jul 1899, *Shear 5625* (HT: US!; CT: NY!).

P. vallicola NAF **24:** 343. 1929. **CALIFORNIA:** San Bernadino Mts., Bear Valley, Jun 1886, *Parish & Parish 1407D* (HT: NY!).

P. zacatecana NAF **24:** 352. 1929. **MEXICO. ZACATECAS:** Casualidad, 26 Apr 1892, *Jones s.n.* (HT: NY!; IT: MO, POM numbered *146,* US!).

Phacomene Amer. J. Bot **16:** 204. 1929. T: *Astragalus beckwithii* Torrey & Gray.

P. pontina NAF **24:** 383. 1929. **CANADA. BRITISH COLUMBIA:** Spence's Bridge, valley of Fraser River, 18 May 1875, *Macoun 450* (HT: GH!).

Phacopsis BTBC **32:** 661. 1905 [1906]. T: *Astragalus praelongus* Sheld.

Pisophaca Amer. J. Bot. **16:** 200. 1929. T: *Phaca flexuosa* Hook.

P. datilensis NAF **24:** 328. 1929. **NEW MEXICO. Socorro Co.:** Negrito Ranger Station, Datil Forest, 22 Sep 1919, *Eggleston 16087* (HT: NY!).

P. ratonensis NAF **24:** 324. 1929. **NEW MEXICO. Union Co.:** Des Moines, 14 Jun 1924, *Eggleston 20114* (HT: NY!; IT: CAS).

P. saundersii NAF **24:** 328. 1929. **COLORADO:** Glenwood Springs, 11 Jun 1893, *Saunders s.n.* (HT: MIN not found by Ownbey in litt. 1984; CT: NY!).

P. sierrae-blancae NAF **24:** 323. 1929. **NEW MEXICO. Lincoln Co.:** White Mts., 26 Jul 1897, *Wooton 213* (HT: NY!; IT: NDG, NMC).

P. stictocarpa NAF **24:** 328. 1929. **NEW MEXICO. Sante Fe Co.:** Santa Fe, 28 May 1897, *Heller & Heller 3606* (HT: NY!; IT: MO, NDG, P, US!).

Psoralea collina Fl. Nebraska **21:** 54. 1895. **NEBRASKA. Davis Co.:** Ft. Robinson, 30 Jul 1887, *Bessey s.n.* (ST: NEB 2 sheets—fide Bolick in litt. 1984); **Scott's Bluff Co.:** Pleasant Valley, 29 Jul 1891, *Rydberg 50* (ST: NEB—fide Bolick in litt. 1984).

P. stenophylla BTBC **40:** 46. 1913. **UTAH. Grand Co.:** Near Wilson Mesa, 1 Jul 1911, *Rydberg & Garrett 8367* (HT: NY!).

P. stenostachys BTBC **40:** 46. 1913. **UTAH. Tooele Co.:** Government Well, 7 Jun 1900, *Jones 6221* (HT: NY!; IT: BRY, CAS, POM, RM!, US).

Psoralidium NAF **24:** 12. 1919. T: *Psoralea tenuiflora* Pursh.

P. batesii Brittonia **1:** 91. 1931. **NEBRASKA:** Wymore, 3 Jul 1909, *Bates 4891* (HT: NY!; IT: NEB).

P. bigelovii NAF **24:** 15. 1919. **NEW MEXICO:** Copper mines, (Santa Rita del Cobre) 1850, *Bigelow 218* (HT: NY!; IT: NY!).

Psorobatus NAF **24:** 40. 1919. T: *Dalea benthamii* T. S. Brandegee.

Psorodendron NAF **24:** 41. 1919. T: *Dalea johnsonii* S. Wats.

Psorothamnus NAF **24:** 45. 1919. T: *Dalea emoryi* A. Gray.

P. dentatus NAF **24:** 47. 1919. **MEXICO. LOWER CALIFORNIA:** Santo Domingo, 26 Sep 1905, *E. W. Nelson & Goldman 7170* (HT: US!; CT: NY!).

P. junceus NAF **24:** 48. 1919. **MEXICO. LOWER CALIFORNIA:** San Pedro Martín Mts., 27 Jun 1905, *Goldman 1178* (HT: US!; CT: NY!).

Pterophacos Fl. Rocky Mts. 507, 1063. 1917 [1918]. T: *Astragalus tetrapterus* A. Gray.

P. cinerascens NAF **24:** 309. 1929. **OREGON. Harney Co.:** Southwest of Narrows, 4 Jul 1912, *Peck 3024* (HT: GH not found).

Rhytidomene NAF **24:** 12. 1919. T: *Psoralea lupinellus* Michx.

Robinia breviloba NAF **24:** 227. 1924. **NEW MEXICO. Grant Co.:** Burrow Mts., 20 Jun 1903, *Metcalfe 189* (HT: NY!).

R. subvelutina NAF **24:** 227. 1924. **ARIZONA:** Natanes Plateau, 26 Jun 1912, *Goodding 1092* (HT: NY!; IT: RM!, US!).

Sauvallella Amer. J. Bot. **11:** 480. 1924. T: *Corynella immarginata* Wright.

Thornbera J. New York Bot. Gard. **20:** 66. 1919. T: *Dalea albiflora* A. Gray.

T. leucantha NAF **22:** 118. 1920. **MEXICO. DURANGO:** Sierra Madre and Sierra Santa Barbara near La Providencia, 11–12 Sep 1898, *E. W. Nelson 4996* (HT: US!; IT: NY!).

T. lutea NAF **24:** 117. 1920. **MEXICO. OAXACA:** Mountain ridge on west side of Valley of Cuicatlán, 10 Nov 1894, *E. W. Nelson 1896* (HT: US!; IT: NY!).

T. nelsonii NAF **24:** 120. 1920. **MEXICO. CHIHUAHUA:** Sierra Madre, 60 miles south of Guadalupe y Calvo, Aug 1898, *E. W. Nelson 4790* (HT: US!; IT: NY! dated 20 Aug).

T. pumila NAF **24:** 120. 1920. **MEXICO. SINALOA:** Culiacán, 5 Oct 1904, *Brandegee s.n.* (HT: US!).

T. villosa NAF **24:** 118. 1920. **ARIZONA:** Santa Rita Mts., 20 Sep–4 Oct 1902, *Griffiths & Thornber 130* (HT: NY!).

Tium egglestonii NAF **24:** 396. 1929. **NEW MEXICO. Socorro Co.:** Datil Forest, 7 Sep 1920, *Eggleston 17216* (HT: NY!; IT: CAS!, US!).

T. platycarpum NAF **24:** 387. 1929. **COLORADO:** Oak Creek, 1873, *Brandegee s.n.* (HT: NY!).

T. sheldoni NAF **24:** 393. 1929. **OREGON. Wallowa Co.:** Horse Creek Canyon, 14 May 1897, *Sheldon 8032* (HT: NY!; IT: MO, US! no data).

T. stenolobum NAF **24:** 388. 1929. **NEW MEXICO.** Sandia Mts., 2 Aug 1914, *Ellis 326* (HT: US!; IT: NY!).

T. variegatum BTBC **34:** 47. 1907. **COLORADO:** Platte Canyon, 19 May 1894, *collector unknown* (from Herb. State Agric. College, Colo.) *15* (HT: NY!).

Trifolium aitonii BTBC **34:** 46. 1907. **IDAHO. Palouse Co.:** Jun–Jul 1892, *Aiton 65* (HT: NY!).

T. confusum BTBC **34:** 46. 1907. **UTAH:** Southern Utah, 1874, *Parry 35* (HT: NY!; IT: NY!).

T. inaequale BTBC **34:** 47. 1907. **UTAH:** Bear River Canyon, Aug 1869, *Watson 243* (HT: NY!; IT: YU!).

T. lilacinum BTBC **28:** 37. 1901. **COLORADO:** West Spanish Peak, 6 Jul 1900, *Rydberg & Vreeland 5950* (HT: NY!; IT: CS, RM!).

T. lividum BTBC **30:** 254. 1903. **COLORADO:** Graymount, 23 Aug 1895, *Rydberg 2442* (HT: NY!; IT: US!).

T. montanense Mem. New York Bot. Gard. **1:** 236. 1900. **MONTANA:** Pony Mts., Old Hollowtop, 9 Jul 1897, *Rydberg & Bessey 4461* (HT: NY!; IT: CAN, MONT, NEB 2 sheets, NY! 2 sheets, RM!, US!).

T. oxyodon E. L. Greene ex Rydb., Fl. Colorado 200–202. 1906. **COLORADO:** Gunnison, 16 Aug 1901, *Baker 826* (HT: NY!).

T. pedunculatum BTBC **30:** 254. 1903. **IDAHO:** Long Valley, 5 Jul 1895, *Henderson 3096* (HT: US!; CT: NY!).

T. salictorum E. L. Greene ex Rydb., Fl. Colorado 200–202. 1906. **COLORADO:** Carson, 2 Jul 1901, *Baker 307* (HT: NY!; IT: DS!, RM! 2 sheets).

T. stenolobum BTBC **28:** 499. 1901. **COLORADO:** La Plata Mts., 15 Jul 1896, *Tweedy 457* (HT: US!; IT: YU!; CT: NY!).

T. uintense BTBC **34:** 47. 1907. **UTAH:** Uintas, Aug 1869, *Watson 241* (HT: NY!; IT: UC, US!, YU!).

Vicia producta BTBC **28:** 500. 1901. **COLORADO:** Butte 5 miles southwest of La Veta, 22 May 1900, *Rydberg & Vreeland 6006* (HT: NY! on 2 sheets; IT: CS, NEB, RM! 2 sheets).

Xylophacos Rydb. *in* Small, Fl. S.E. U.S. 619, 1332. 1903. T: *Astragalus shortianus* Nutt.

X. aragalloides BTBC **34:** 48. 1907. **UTAH:** St. George, 5 Apr 1880, *Jones 1633* (HT: NY!; IT: GH, ISC, POM, UT!).

X. argentinus BTBC **52:** 371. 1925. **CALIFORNIA:** Lone Pine, 14 May 1897, *Jones s.n.* (HT: GH; IT: US!).

X. blyae NAF **24:** 303. 1929. **ARIZONA:** 1½ miles northwest of Kingman, 1927, *Bly s.n.* (HT: US!; IT: NY!).

X. incurvus BTBC **52:** 366. 1925. **CALIFORNIA:** 1875, *Lemmon 76* (HT: NY!).

X. iodopetalus BTBC **52:** 152. 1925. **COLORADO:** Arboles, Jun 1899, *Baker 415* (HT: NY!; IT: GH, K, MO, NDG, P, RM! 2 sheets, US!).

X. lagopinus BTBC **52:** 372. 1925. **OREGON. Crook Co.:** Between Preneville and Bear Buttes, 25 Jun 1894, *Leiberg 326* (HT: NY!; IT: MIN, POM, US!).

X. lenophyllus Fl. Rocky Mts. ed. 2, 1126. 1922 [1923]. **ARIZONA:** Vicinity of Flagstaff, 2 Jun 1898, *MacDougal 27* (HT: NY!; IT: GH, POM).

X. marianus BTBC **52:** 233. 1925. **UTAH:** Marysvale, 1 Jun 1894, *Jones 5355* (HT: NY!; IT: MO, POM!, US!).

X. medius BTBC **52:** 232. 1925. **UTAH:** Lake Point, 29 May 1880, *Jones 1743* (HT: NY!; IT: BRY, CAS!, DS! 2 sheets, POM, US!, UT!).

X. melanocalyx BTBC **52:** 149. 1925. **UTAH:** Copper mine west of St. George, 4 Apr 1894, *Jones 5009* (HT: POM; IT: NY! labeled Beaverdam, US! labeled 12 miles north of Beaverdam).

X. subvillosus BTBC **52:** 368. 1925. **CALIFORNIA:** Sierra Valley, 1873, *Lemmon s.n.* (HT: NY! not dated; IT: GH!).

X. tidestromii BTBC **52:** 155. 1925. **NEVADA:** Charleston Mts., 27 May 1919, *Tidestrom 9661* (HT: US!; CT: NY!).

X. ventosus BTBC **52:** 370. 1925. **WASHINGTON. Klickitat Co.:** Bingen, 10 Nov 1920, *Suksdorf 10662* (HT: NY!; IT: WS).

FAGACEAE

Quercus eastwoodiae Bull. New York Bot. Gard. **2:** 210. 1901. **UTAH. San Juan Co.:** Butler Wash, 17 Jul 1895, *Eastwood 141* (HT: NY!; IT: US!).

Q. havardii Bull. New York Bot. Gard. **2:** 213. 1901. **TEXAS:** Sandhills of staked plains, *Havard 51* (HT: MO—fide Wolf in litt. 1984; IT: US! dated Sep 1881 and no collector listed).

Q. leptophylla Bull. New York Bot. Gard. **2:** 205. 1901. **COLORADO:** Tributaries of Turkey Creek, 8–9 Jun 1900, *Rydberg & Vreeland 6347* (HT: NY!; IT: NEB, RM!).

Q. mohriana Buckley ex Rydb., Bull. New York Bot. Gard. **2:** 219. 1901. **TEXAS:** Valley of Devil's River, 1875, *Buckley s.n.* (HT: NY!; IT: NY!).

Q. nitescens Bull. New York Bot. Gard. **2:** 207. 1901. **COLORADO:** Near La Veta, 18 Sep 1900, *Vreeland 677* (HT: NY!).

Q. pauciloba Bull. New York Bot. Gard. **2:** 215. 1901.

ARIZONA: Beaver Creek, 7 Aug 1891, *MacDougal s.n.* (HT: US!; IT: NY!).

Q. submollis Bull. New York Bot. Gard. **2:** 202. 1901. ARIZONA: Fort Huachuca, Sep 1894, *Wilcox 491* in part (HT: US!; IT: NY!).

Q. vreelandii Bull. New York Bot. Gard. **2:** 204. 1901. COLORADO: Near La Veta, 20 Sep 1900, *Vreeland 685* (HT: NY!; IT: RM! dated 26 Sep).

Q. wilcoxii Bull. New York Bot. Gard. **2:** 227. 1901. ARIZONA: Fort Huachuca, Jul 1892, *Wilcox s.n.* (HT: NY!).

FUMARIACEAE

Bicuculla occidentalis BTBC **29:** 160. 1902. WASHINGTON. **West Klickitat Co.:** 11 Apr 1892, *Suksdorf 1946* (HT: NY!; IT: US!).

Capnoides brachycarpum BTBC **34:** 426. 1907. UTAH: Alta, 5 Aug 1879, *Jones 1197* (HT: NY!; IT: BRY, NY!, POM, RM!, US!, UTC!).

C. hastatum BTBC **34:** 426. 1907. IDAHO: Lolo Creek, 15 Jul 1902, *Piper 4057* (HT: NY!).

GENTIANACEAE

Amarella ventorum BTBC **40:** 463. 1913. WYOMING: Wind River Mts., 4 Aug 1843, *Fremont s.n.* (HT: NY not found).

Dasystephana obtusiloba BTBC **40:** 464. 1913. MONTANA: Mary Baker Lake and Sperry Glacier, 21 Aug 1901, *Vreeland 1162* (HT: NY!; IT: US!).

Frasera speciosa var. *angustifolia* BTBC **31:** 632. 1904. [1905]. MONTANA: Lima, 30 Jun 1895, *Shear 3369* (HT: NY!).

F. speciosa var. *stenosepala* BTBC **31:** 632. 1904 [1905]. COLORADO. **Larimer Co.:** 20 Jun 1895, *Cowan s.n.* (HT: NY!).

Gentianella clementis BTBC **31:** 631. 1904 [1905]. COLORADO: Minnehaha, 31 Jul 1901, *Clements & Clements 253* (HT: NY!; IT: NEB, RM!, US!).

Leucocraspedum Fl. Rocky Mts. 665, 1065. 1917 [1918]. T: *Frasera albomarginata* S. Wats.

Swertia fritillaria BTBC **40:** 465. 1913. UTAH: Big Cottonwood Canyon, 4 Aug 1905, *Garrett 1566* (HT: NY!; IT: UT!).

GERANIACEAE

Geranium albidum Rydb. ex Hanks & Small, NAF **25:** 19. 1907. **MEXICO.** CHIHUAHUA: Near Colonia García in the Sierra Madres, 29 Jul 1899, *Townsend & Barber 181* (HT: NY!; IT: US!).

G. cowanii [*cowenii* in publication] Fl. Colorado 218. 1906. COLORADO: Rist Canyon, 16 Jun 1899, *Marshall 1157* (LT: NY!—Jones & Jones, Rhodora **45:** 42. 1943; ILT: CS, NY!).

G. deltoideum Rydb. ex Hanks & Small, NAF **25:** 18. 1907. **MEXICO.** DURANGO: San Ramón, 21 Apr–18 May 1906, *Palmer 192* (HT: NY!; IT: US!).

G. gracilentum E. L. Greene ex Rydb., Fl. Colorado 218. 1906. COLORADO: Piedra, 14 Jul 1899, *Baker 449* (HT: NY!; IT: US!).

G. marginale Rydb. ex Hanks & Small, NAF **25:** 16. 1907. UTAH: Aquarius Plateau, at the head of Poison Creek, 4 Aug 1905, *Rydberg & Carlton 7401* (HT: NY!).

G. nervosum BTBC **28:** 34. 1901. WYOMING: Fish Creek, Teton Forest Reserve, Jul 1897, *Tweedy 494* (HT: NY!—designated by Rydberg in herb.).

G. pattersonii BTBC **29:** 242. 1902. COLORADO: Gray's Peak, 23 Aug 1895, *Rydberg s.n.* (HT: NY!).

G. regale Rydb. ex Hanks & Small, NAF **25:** 11. 1907. **MEXICO.** HIDALGO: Near Real del Monte, 19 Jul 1905, *Rose, Painter & Rose 8675* (HT: NY!; IT: US!).

G. strigosum BTBC **29:** 243. 1902. WYOMING. **Carbon Co.:** Coppertown, 20 Jun 1901, *Tweedy 4591* (HT: NY!; IT: US!, YU!).

G. thermale Mem. New York Bot. Gard. **1:** 478. 1900. MONTANA: Lo-Lo Hot Springs, 1898, *Williams & Griffith s.n.* (HT: NY!).

GROSSULARIACEAE

Limnobotrya Fl. Rocky Mts. 396, 1062. 1917 [1918]. T: *Ribes oxyacanthoides* var. *lacustre* Pers.

Ribes aureum var. *chrysococcus* Fl. Nebraska **21:** 71. 1895. NEBRASKA: Scott's Bluff, 21 Jul 1891, *Rydberg 106b* (ST: NEB—fide Bolick in litt. 1985, US!); **Hooker Co.:** Middle Loup River near Mullen, 27 Jul 1893, *Rydberg 1601* (ST: NEB—fide Bolick in litt. 1985).

R. huronense Rydb. *in* Britton, Man. fl. n. states 487. 1901. Lake Huron, *Todd s.n.* (HT: K).

R. inerme Mem. New York Bot. Gard. **1:** 202. 1900. WYOMING: Yellowstone Park, Slough Creek, Jul 1885, *Tweedy 830* (HT: YU!; IT: NY! dated Jun).

R. vallicola E. L. Greene ex Rydb., Fl. Colorado 176, 177. 1906. COLORADO: Los Pinos (Bayfield), 27 May 1899, *Baker 372* (HT: NY!; IT: MO, NDG, POM).

HYDRANGEACEAE

Philadelphus angustifolius Rydb. *in* Small & Rydb., NAF **22:** 166. 1905. WASHINGTON: Palace Camp, 1883, *Willis s.n.* (HT: NY!; IT: NY!).

P. argenteus Rydb. *in* Small & Rydb., NAF **22:** 171. 1905. ARIZONA: Ft. Huachuca, Jul 1893, *Wilcox s.n.* (HT: NY!; IT: A).

P. ellipticus Rydb. *in* Small & Rydb., NAF **22:** 172. 1905. NEW MEXICO: Mesilla Park, 1896, *Tinsley s.n.* (HT: US!; CT: NY!).

P. fremontii Rydb. *in* Small & Rydb., NAF **22:** 165. 1905. CALIFORNIA: American River, 1846, *Fremont 511* (HT: NY!).

P. helleri Rydb. *in* Small & Rydb., NAF **22:** 166. 1905. IDAHO. **Nez Perces Co.:** Lake Waha, 4 Jul 1896, *Heller & Heller 3374* (HT: NY!; IT: CAS, NDG, NY! numbered *3474,* US!).

P. mearnsii W. H. Evans ex Rydb. *in* Small & Rydb., NAF **22:** 174. 1905. NEW MEXICO. **Grant Co.:** 5 May 1892, *Mearns 36* (HT: US; IT: NY! dated 28 Apr).

P. minutus Rydb. *in* Small & Rydb., NAF **22:** 173.

1905. COLORADO: Black Canyon of the Gunnison, 27 Jun 1901, *Baker 266* (HT: NY!; IT: GH, K, NDG, POM, RM, US!).

P. palmeri Rydb. *in* Small & Rydb., NAF **22:** 173. 1905. **MEXICO. [COAHUILA]:** Sierra Madre, 40 miles south of Saltillo, Jul 1880, *Palmer 2122* (HT: GH!).

P. platyphyllus Rydb. *in* Small & Rydb., NAF **22:** 167. 1905. CALIFORNIA. **Siskiyou Co.:** Mt. Shasta, 1–15 Jul 1897, *Brown 562* (HT: NY!; IT: US!).

P. pumilus Rydb. *in* Small & Rydb., NAF **22:** 173. 1905. CALIFORNIA: San Jacinto Mts., Tamarack Valley, Jul–Aug 1901, *Hall 2500* (HT: NY!; IT: GH, MO, US!).

P. stramineus Rydb. *in* Small & Rydb., NAF **22:** 172. 1905. CALIFORNIA. **Mono Co.:** White Mts., Aug 1888, *Shockley s.n.* (HT: NY!; IT: NY!).

HYDROCHARITACEAE

Philotria linearis BTBC **35:** 464. 1908. TENNESSEE: Cumberland River, vicinity of Nashville, *Gattinger s.n.* (HT: NY!; IT: DS!, US! just labeled Nashville).

HYDROPHYLLACEAE

Phacelia alba BTBC **28:** 30. 1901. COLORADO: Sangre de Cristo Creek, 2 Jul 1900, *Rydberg & Vreeland 5755* (HT: NY!).

P. alpina Mem. New York Bot. Gard. **1:** 324. 1900. MONTANA: Cedar Mt., 16 Jul 1897, *Rydberg & Bessey 4855* (HT: NY!; IT: US! dated Jul 1897).

P. burkei BTBC **36:** 675. 1909. IDAHO: Snake Country, *Burke s.n.* (HT: NY!).

P. leptosepala BTBC **36:** 676. 1909. CANADA. BRITISH COLUMBIA: Vermilion Lake, 5 Aug 1905, *Farr 1013* (HT: NY!).

P. nervosa BTBC **36:** 675. 1909. COLORADO: Silver Plume, 24 Aug 1895, *Rydberg s.n.* (HT: NY! on 2 sheets).

P. orbicularis BTBC **40:** 479. 1913. UTAH: Marvine Laccelite, 2 Jul 1894, *Jones 5663am* (HT: US!; CT: NY!).

P. sericea var. *ciliosa* BTBC **31:** 636. 1904 [1905]. COLORADO: **Rio Blanco Co.:** North of Meeker, 9 Jun 1902, *Osterhout 2619* (HT: NY!).

HYPERICACEAE

Hypericum mitchellianum Torreya **27:** 84. 1927. NORTH CAROLINA: Mt. Mitchell, 15–16 Jul 1925, *Rydberg 9402* (HT: NY!).

JUNCACEAE

Juncoides subcapitatum BTBC **31:** 401. 1904. COLORADO: Silver Plume, 24 Aug 1895, *Shear 4611* (HT: NY!).

Juncus balticus var. *vallicola* BTBC **31:** 399. 1904. WYOMING: Point of Rocks, 17 Jun 1901, *Merrill & Wilcox 664* (HT: NY!).

J. brunnescens BTBC **31:** 400. 1904. COLORADO: Pagosa Springs, Jul 1899, *Baker 245* (HT: NY!; IT: US!).

J. jonesii Fl. Rocky Mts. 153, 1061. 1917 [1918]. UTAH: Wasatch Mts., Alta, 6 Aug 1879, *Jones s.n.* (HT: NY!; IT: NY! 3 sheets, POM).

J. parous BTBC **31:** 401. 1904. COLORADO: [Ft.] Garland, 17 Jul 1896, *Shear 3666* (HT: NY!).

J. suksdorfii BTBC **26:** 541. 1899. WASHINGTON: Falcon Valley, 30 Jun 1883, *Suksdorf 217* (ST: GH!); same locality, 1 Aug 1885, *Suksdorf 680* (ST: GH!); **Spokane Co.:** Spangle, 27 Jun 1884, *Suksdorf 464* (ST: GH!).

J. tracyi Fl. Rocky Mts. 155, 1061. 1917 [1918]. UTAH: Ogden, 2 Aug 1887, *Tracy 389* (HT: NY!; CT: US).

J. truncatus BTBC **31:** 399. 1904. COLORADO: Meadow Height, 17 Aug 1898, *Shear & Bessey 4323* (HT: NY!).

J. tweedyi Mem. New York Bot. Gard. **1:** 90. 1900. WYOMING: Yellowstone Park, Aug 1884, *Tweedy 223* [perhaps *225*] (HT: YU!).

LAMIACEAE

Audibertiella argentea BTBC **36:** 683. 1909. ARIZONA: Mokiak Pass, 1877, *Palmer 395* (HT: NY!; IT: NY! 3 sheets).

Hedeoma longiflora BTBC **36:** 685. 1909. NEBRASKA. **Banner Co.:** Aug 1890, *Rydberg 297* (HT: NY!; IT: NEB).

Lycopus velutinus Rydb. ex Small, Fl. S.E. U.S. 1049, 1337. 1903. TEXAS: Tyler, 20 Sep 1900, *Reverchon 2120* (HT: NY!).

Madronella oblongifolia BTBC **36:** 686. 1909. UTAH: Mt. Nebo, 15 Aug 1905, *Rydberg & Carlton 7706* (HT: NY!; IT: RM, US!).

M. sessilifolia BTBC **36:** 685. 1909. UTAH: St. George, 1877, *Palmer 393* (HT: NY!).

Mentha occidentalis BTBC **36:** 687. 1909. IDAHO. **Nez Perces Co.:** 30 Jul 1896, *Heller & Heller 3486* (HT: NY!).

M. rubella Mem. New York Bot. Gard. **1:** 337. 1900. WYOMING: Yellowstone Park, lower Geyser Basin, 4 Aug 1897, *Rydberg & Bessey 4900* (HT: NY! on 2 sheets; IT: MONT, NEB 2 sheets, US!).

Monarda comata BTBC **28:** 502. 1901. COLORADO: Wahatoya Creek, 7 Jul 1900, *Rydberg & Vreeland 5673* (HT: NY!).

Monardella dentata BTBC **31:** 641. 1904. [1905]. COLORADO: Gray's Peak, Aug & Sep 1872, *Torrey s.n.* (HT: NY!).

Scutellaria veronicifolia BTBC **36:** 681. 1909. IDAHO. **Nez Perces Co.:** Valley of Peter Creek, 4 May 1892, *Sandberg, MacDougal & Heller 115* (HT: NY!).

Stachys ampla BTBC **36:** 683. 1909. SOUTH DAKOTA: Black Hills, Custer, 20 Aug 1892, *Rydberg 1208* (HT: NY!).

S. asperrima BTBC **36:** 682. 1909. UTAH: Towards Jordan City, 7 Jul 1884, *Leonard 138* (HT: NY!; IT: NY!).

S. borealis Brittonia **1:** 95. 1931. **CANADA.** Mackenzie Territory, Mackenzie River, 1861–1862, *Onion, Kennicott & Hardisty s.n.* (HT: NY!).

S. brevidens Brittonia **1:** 95. 1931. **Minnesota:** Tower, 1891, *Lugger s.n.* (HT: MIN—fide Ownbey in litt. 1984).

S. leibergii BTBC **36:** 682. 1909. **Idaho:** Coeur d'Alene Mts., Blue Creek, 20 Jul 1895, *Leiberg 1328* (HT: US!; IT: NY!).

S. pustulosa Brittonia **1:** 95. 1931. **Wisconsin:** Fort Howard, 22 Aug 1890, *Schuette s.n.* (HT: NY!).

S. subcordata Torreya **27:** 88. 1927. **Virginia. Bedford Co.:** Alleghany Mts., peaks of Otter, 1 Jul 1925, *Rydberg 9264* (HT: NY! on 2 sheets; IT: CAS!, US! dated 1–2 Jul).

S. teucriformis BTBC **31:** 640. 1904 [1905]. **Colorado. Weld Co.:** New Windsor, 5 Jul 1894, *Osterhout s.n.* (HT: NY!).

LILIACEAE

Allium aridum Fl. Rocky Mts. 159, 1061. 1917 [1918]. **Wyoming:** Green River, 25 Jun 1895, *Rydberg 2605* (HT: NY!; IT: NY!).

A. fibrosum BTBC **24:** 188. 1897. **Montana:** Near Lima, 29 Jun 1895, *Rydberg 2606* (HT: NY!).

A. macropetalum BTBC **31:** 401. 1904. **Colorado:** 14 May 1898, *Crandall 4082* (HT: NY!; IT: CS labeled Palisades).

A. neo-mexicanum BTBC **26:** 541. 1899. **New Mexico:** Organ Mts., 29 Aug 1894, *Wooton s.n.* (HT: NY!—designated by Rydberg in herb.).

A. pikeanum BTBC **31:** 402. 1904. **Colorado:** Pike's Peak, Halfway House, 3 Jul 1896, *Shear 3721* (HT: NY!; IT: US!).

A. recurvatum Mem. New York Bot. Gard. **1:** 94. 1900. **Montana:** Indian Creek, 2 Jul 1897, *Rydberg & Bessey 3850* (HT: NY!).

Anticlea vaginata BTBC **39:** 108. 1912. **Utah:** Armstrong Canyon near natural bridges, 4–6 Aug 1911, *Rydberg & Garrett 9407* (HT: NY!; IT: BRY, CAN, NY!, RM, US!, UT!).

Calochortus acuminatus BTBC **24:** 189. 1897. **Montana:** Lima, 5 Aug 1895, *Rydberg 2600* (HT: NY!).

Dipterostemon BTBC **39:** 110. 1912. Ts: *Brodiaea capitata* Benth., *B. insularis* E. L. Greene, *B. capitata* var. *pauciflora* Torrey, *Hookera pulchella* Salisb.

Erythronium leptopetalum Fl. Rocky Mts. 165, 1061. 1917 [1918]. **Idaho:** Near Boise, Jun 1892, *Mulford s.n.* (HT: NY!; IT: NEB).

E. utahense Fl. Rocky Mts. 165, 1061. 1917 [1918]. **Utah:** Salt Lake City, 1850, *Stansbury s.n.* (HT: NY!).

Ochrocodon Fl. Rocky Mts. 164, 1061. 1917 [1918]. T: *Lilium pudicum* Pursh.

Stenanthella BTBC **27:** 530. 1900. T: *Stenanthium occidentale* A. Gray.

Streptopus curvipes Vail ex Rydb., BTBC **28:** 267. 1901. **CANADA. British Columbia:** Asulkon Pass, Jul 1897, *Z. Palmer s.n.* (HT: NY!).

Tofieldia intermedia BTBC **27:** 528. 1900. **Alaska:** Sheh-Shooh Lake, 6 Jul 1895, *Gorman 78* (HT: NY!; IT: US!).

Toxicoscordion BTBC **30:** 272. 1903. T: *Zigadenus intermedius* Rydb.

T. texense Rydb. ex Small, Fl. S.E. U.S. 252, 1328. 1903. **Texas:** *Drummond 420* (HT: NY!).

Vagnera australis Rydb. ex Small, Fl. S.E. U.S. 270, 1328. 1903. **Alabama. See Co.:** Auburn, 22 Apr 1896, *Earle & Underwood s.n.* (HT: NY!).

V. brachypetala BTBC **28:** 268. 1901. **CANADA. British Columbia:** Glacier at the "Loup," Aug 1897, *Van Brunt & Van Brunt s.n.* (HT: NY!).

V. leptopetala BTBC **28:** 268. 1901. **Colorado:** Headwaters of Sangre de Cristo Creek, 22 Jun 1900, *Rydberg & Vreeland 6441* (HT: NY!).

Veratrum speciosum BTBC **27:** 531. 1900. **Montana:** Bridger Mts., 28 Jul 1896, *Flodman 344* (HT: NY!; IT: US!).

Zigadenus acutus BTBC **27:** 536. 1900. **South Dakota:** Black Hills, Box Elder Creek, Jun 1887, *W. Rusby s.n.* (HT: NY!).

Z. coloradensis BTBC **27:** 534. 1900. **Colorado:** Idaho Springs, Aug 1895, *Rydberg s.n.* (HT: NY!).

Z. falcatus BTBC **27:** 536. 1900. **Colorado:** Fort Collins, 23 May 1893, *Crandall s.n.* (HT: NY!; IT: CS, US! numbered *487*).

Z. gramineus BTBC **27:** 535. 1900. **Montana. Gallatin Co.:** Spanish Basin, 28 Jun 1897, *Rydberg & Bessey 3848* (HT: NY!).

Z. intermedius BTBC **27:** 535. 1900. **Idaho. Nez Perces Co.:** May 1892, *Sandberg 10564* (HT: NY!).

LIMNANTHACEAE

Floerkia occidentalis Mem. New York Bot. Gard. **1:** 268. 1900. **Utah:** Wasatch Mts., Jun 1869, *Watson 208* (ST: NY!). **Washington Territory:** 1838–1842, *Wilkes Expedition s.n.* (ST: NY!, US! labeled upper Columbia and numbered *837*). **Wyoming:** Yellowstone Park, Swan Lake, Jun 1885, *Tweedy 525* (ST: US!, YU!).

LOASACEAE

Acrolasia gracilis BTBC **31:** 566. 1904. **Colorado. Larimer Co.:** 1 Jun 1895, *Cowan s.n.* (HT: NY!).

A. latifolia BTBC **31:** 567. 1904. **Colorado:** Mountains between Sunshine and Ward, Aug 1902, *Tweedy 5149* (HT: NY!).

Mentzelia ctenophora BTBC **28:** 33. 1901. **Colorado:** Cuchara River below La Veta, 30 May 1900, *Rydberg & Vreeland 5769* (HT: NY!; IT: NEB).

M. tenerrima Mem. New York Bot. Gard. **1:** 271. 1900. **Montana:** Electric Peak, 18 Aug 1897, *Rydberg & Bessey 4542* (HT: NY! on 2 sheets; IT: CAN, MONT, NEB, US!).

M. tweedyi Mem. New York Bot. Gard. **1:** 271. 1900. **Montana. Park Co.:** Trail Creek, 13 Jul 1887, *Tweedy 152* (HT: YU!).

Nuttallia acuminata BTBC **40:** 61. 1913. **Idaho. Koo-**

tenai Co.: Spokane River, 15 Jul 1892, *Sandberg, MacDougal & Heller 651* (HT: NY!).

N. lobata BTBC **40:** 61. 1913. **Utah:** Southern Utah, 1877, *Palmer 172* (HT: NY!; IT: NY!).

Touterea laciniata BTBC **31:** 565. 1904. **Colorado:** Pagosa Springs, Jul 1899, *Baker 470* (HT: NY!; IT: US!).

T. sinuata BTBC **31:** 566. 1904. **Colorado:** Boulder, 2 Sep 1895, *Rydberg s.n.* (HT: NY!; IT: NY!).

MALVACEAE

Sphaeralcea arizonica A. A. Heller ex Rydb., BTBC **40:** 59. 1913. **Arizona:** Flagstaff, 16 Jun 1898, *MacDougal 120* (HT: NY!; IT: NY!, US!).

S. crandallii BTBC **31:** 564. 1904. **Colorado:** Mountains west of Steamboat Springs, 14 Jul 1894, *Crandall 97* (HT: NY not found).

S. grandiflora BTBC **31:** 565. 1904. **Colorado:** Mesa Verde, 14 Jun 1901, *Vreeland 882* (HT: NY!; IT: NY!).

S. marginata York ex Rydb., BTBC **33:** 145. 1906. **Colorado:** Grand Junction, 11 Jun 1901, *Baker 93* (HT: NY!; IT: US!).

S. subrhomboidea BTBC **40:** 59. 1913. **Utah: Wasatch Co.:** Near Midway, 6 Jul 1905, *Carlton & Garrett 6691* (HT: NY! on 2 sheets; IT: US!).

MONOTROPACEAE

Hypopitys latisquama BTBC **40:** 461. 1913. **Montana:** Bridger Mts., 28 Jul 1896, *Flodman 708* (HT: NY!; IT: NY!).

NYCTAGINACEAE

Abronia arenaria Mem. New York Bot. Gard. **1:** 137. 1900. **Wyoming:** Yellowstone Park, mouth of Pelican Creek, Aug 1885, *Tweedy 442* (HT: YU!; IT: NY!).

A. fallax Heimerl ex Rydb., BTBC **29:** 684. 1902. **Utah:** Salt Lake City, 26 Aug 1879, *Jones 1337* (HT: US!; IT: NY! 2 sheets, UTC).

A. glabra BTBC **29:** 685. 1902. **Colorado:** Grand Junction, 9 Jul 1883, *Jones s.n.* (HT: US!).

A. lanceolata BTBC **29:** 685. 1902. **Idaho:** Idaho Falls, 4 Jul 1901, *Merrill & Wilcox 870* (HT: NY!; IT: US!).

A. nudata BTBC **29:** 683. 1902. **Montana: Dawson Co.:** Colgate near Glendive, 6 Sep 1892, *Sandberg, MacDougal & Heller 1016* (HT: NY!; IT: DS!, US!).

A. pumila BTBC **29:** 683. 1902. **Utah:** Emery, 16 Jun 1894, *Jones 5445q* (HT: US!; IT: POM; CT: NY!).

A. salsa BTBC **29:** 684. 1902. **Utah:** Salt Lake City, May 1869, *Watson 965* (HT: NY!; IT: GH!, US!, YU!).

Allionia bracteata BTBC **29:** 690. 1902. **Missouri:** Malden, 21 Aug 1894, *Bush 459* (HT: NY!; IT: NY!).

A. divaricata BTBC **29:** 691. 1902. **Colorado:** Durango, 18 Jul 1898, *Baker, Earle & Tracy 512* (HT: NY!).

A. lanceolata BTBC **29:** 691. 1902. **Colorado. Larimer Co.:** Estes Park, 10 Jul 1897, *Osterhout 1556* (HT: NY!).

Allioniella BTBC **29:** 687. 1902. T: *Quamoclidion oxybaphoides* A. Gray. (HT: NY!; IT: GH!, US!, YU!).

ONAGRACEAE

Anogra cinerea BTBC **31:** 570. 1904. **Colorado:** Between Bent's Fort and Pueblo, 1845, *Fremont 234* (HT: NY!).

A. perplexa Fl. Rocky Mts. 596, 1064. 1917 [1918]. **Colorado:** Near Dolores, 12 Jun 1901, *Vreeland 880* (HT: NY!; IT: NY!).

A. vreelandii BTBC **31:** 570. 1904. **Colorado:** McElmo Canyon, 10 Jun 1901, *Vreeland 861* (HT: NY!; IT: NY!).

Boisduvalia salicina BTBC **40:** 62. 1913. **Oregon:** *Nuttall s.n.* (HT: NY!).

Epilobium drummondii var. *latiusculum* Mem. New York Bot. Gard. **1:** 276. 1900. **Idaho:** Henry's Lake and Mt. Chauvet, 31 Jul 1897, *Rydberg & Bessey 4559* (HT: NY!—designated by Rydberg in herb.; IT: MONT, NEB, US).

E. laevicaule BTBC **40:** 64. 1913. **Montana:** Manhattan, 24 Jul 1895, *Rydberg 2728* (HT: NY!; IT: NY!).

E. macdougalii Fl. Rocky Mts. 587, 1064. 1917 [1918]. **Montana:** Eastern shore of Flathead Lake, 29 Jul 1901, *MacDougal 793* (HT: NY!).

E. ovatifolium BTBC **31:** 567. 1904. **Colorado:** Near Empire, 10 Aug 1892, *Patterson 205* in part (HT: NY!).

E. palmeri BTBC **31:** 569. 1904. **Utah:** Southern Utah, *Palmer 156* (HT: NY!).

E. rubescens BTBC **31:** 568. 1904. **Colorado:** Pagosa Springs, Jul 1899, *Baker s.n.* (HT: NY!).

E. sandbergii BTBC **40:** 64. 1913. **Idaho. Kootenai Co.:** Valley of Mud Lake, 25 Jul 1892, *Sandberg, MacDougal & Heller 737* (HT: NY!; IT: DS!, US!).

E. stramineum BTBC **31:** 568. 1904. **Colorado:** Idaho Springs, 26 Aug 1895 [1905 in publication], *Rydberg s.n.* (HT: NY!).

E. tracyi BTBC **40:** 63. 1913. **Utah:** Ogden, 31 Jul 1887, *Tracy & Evans 547* (HT: NY!).

Gaura coloradensis BTBC **31:** 572. 1904. **Colorado:** Fort Collins, 8 Jul 1895, *Cowan s.n.* (HT: NY!; IT: US! dated 1 Jul).

Gayophytum helleri BTBC **40:** 65. 1913. **Idaho. Nez Perces Co.:** 16 Jul 1896, *Heller & Heller 3433* (HT: NY!; IT: NY!, US!).

G. intermedium BTBC **31:** 569. 1904. **Colorado:** Ouray, 9 Sep 1901, *Underwood & Selby 193* (HT: NY!).

Kneiffia latifolia Torreya **27:** 86. 1927. **North Carolina. Buncombe Co.:** Craggy Mts., 21 Jul 1925, *Rydberg 9455* (HT: NY!).

Meriolix intermedia Rydb. ex Small, Fl. S.E. U.S. 846, 1335. 1903. **Missouri:** Watson, 1 Jun 1894, *Bush 321* (HT: NY!).

M. melanoglottis Rydb. ex Small, Fl. S.E. U.S. 846, 1335. 1903. **Texas. Kerr Co.:** Kerrville, 19–25 Apr 1894, *Heller 1600* (HT: NY!).

M. oblanceolata Brittonia **1:** 93. 1931. **Kansas:** Vicinity of Coldwater, 2 miles west of town, 8 Jul 1929, *Rydberg & Imler 737* (HT: NY! on 2 sheets; IT: US!).

Oenothera longissima BTBC **40:** 65. 1913. **Utah:** Armstrong and White canyons near the natural bridges, 4–6 Aug 1911, *Rydberg & Garrett 9410* (HT: NY!; IT: US!).

O. pallida var. *latifolia* Contr. U.S. Natl. Herb. **3:** 159. 1895. **Nebraska. Hooker Co.:** Forks of Dismal River, 17 Jul 1893, *Rydberg 1544* (LT: US—Munz, Amer. J. Bot. **18:** 322. 1931; ILT: NY! labeled middle Loup River).

Onagra strigosa Mem. New York Bot. Gard. **1:** 278. 1900. **Montana:** Old Hollowtop near Pony, 12 Jul 1897, *Rydberg & Bessey 4584* (HT: NY!—designated by Rydberg in herb.; IT: CAN, MONT, NEB 2 sheets, US! 2 sheets).

O. strigosa var. *subulata* Mem. New York Bot. Gard. **1:** 279. 1900. **Montana:** Forks of the Madison, 26 Jul 1897, *Rydberg & Bessey 4588* (HT: NY!; IT: US!).

Pachylophus caulescens BTBC **31:** 571. 1904. **Colorado:** Palisades, bank of the Grand River, 29 May 1894, *Crandall 12* (HT: NY!; probable IT: CS 2 sheets).

P. crinitus Fl. Rocky Mts. 598, 1064. 1917 [1918]. **Utah:** Rabbit Valley, 6 Aug 1875, *Ward 526* in part (HT: US!).

P. hirsutus BTBC **31:** 571. 1904. **Colorado:** Georgetown, 18 Aug 1895, *Rydberg s.n.* (HT: NY!).

P. macroglottis BTBC **30:** 259. 1903. **Colorado:** Turkey Creek and tributaries, 8–9 Jun 1900, *Rydberg & Vreeland 5857* (HT: NY!; IT: RM).

Zauschneria crassifolia Fl. Rocky Mts. 590, 1064. 1917 [1918]. **Arizona:** D. A. Adams Ranch, 8 Oct 1900, *Griffiths 1842* (HT: NY!).

ORCHIDACEAE

Aplectrum shortii Rydb. *in* Britton, Man. fl. n. states 305. 1901. **Kentucky:** *Short s.n.* (ST: NY!); Lexington, *Short s.n.* (ST: NY!).

Corallorrhiza ochroleuca BTBC **31:** 402. 1904. **Nebraska:** War Bonnet Canyon, 24 Jun 1890, *Williams 48* (HT: NY!; IT: NY!, US! dated Jun 1890 and not numbered).

C. vreelandii BTBC **28:** 271. 1901. **Colorado:** Veta Mt., 4 Jun 1900, *Rydberg & Vreeland 6418* (HT: NY!; IT: NY!).

Denslovia Brittonia **1:** 85. 1931. T: *Orchis clavellata* Michx.

Galeorchis Rydb. *in* Britton, Man. fl. n. states 292. 1901. T: *Orchis spectabilis* L.

Gymnadeniopsis Rydb. *in* Britton, Man. fl. n. states 293. 1901. T: *Orchis nivea* Nutt. (Typified by Britton & Brown, Ill. Fl. N. U.S. & Canada, II, **1:** 552, 1913.)

Gyrostachys linearis Rydb. *in* Britton, Man. fl. n. states 300. 1901. **New Jersey:** Closter, *Austin s.n.* (HT: NY!; IT: NY! 2 sheets one dated 1858).

G. ochroleuca Rydb. *in* Britton, Man. fl. n. states 300. 1901. Typified by Sheviak and Catling (Rhodora **82:** 541. 1980) on a drawing at NY! labeled: *Gyrostachys ochroleuca* Rydberg ined. Britton, Man. fl. n. states, 1901, drawing by Mrs. Long, Mt. Washington, Mass., 1897.

G. stricta Mem. New York Bot. Gard. **1:** 107. 1900. **Montana:** Forks of the Madison, 26 Jul 1897, *Rydberg & Bessey 3908* (HT: NY!—designated by Rydberg in herb.).

Limnorchis Mem. New York Bot. Gard. **1:** 104. 1900. T: *Orchis hyperborea* L.

L. arizonica BTBC **28:** 629. 1901. **Arizona:** 1891, *Nealley s.n.* (HT: NY!).

L. behringiana BTBC **28:** 620. 1901. **[SIBERIA].** Bering Island, Aug 1891, British Bering Sea Commission # *143* (HT: NY!).

L. brachypetala Rydb. *in* Britton & Rydb., Bull. New York Bot. Gard. **2:** 161. 1901. **CANADA. Yukon:** Bennett City, 25 Aug 1899, *Williams s.n.* (HT: NY!).

L. dilatata var. *linearifolia* BTBC **28:** 623. 1901. **New York:** Bridgewater, 15 Jun, *Gray s.n.* (HT: NY!).

L. ensifolia BTBC **28:** 629. 1901. **Arizona:** Mt. Humphrey, 31 Jul 1897, *Kunze s.n.* (HT: NY!).

L. foliosa BTBC **28:** 622. 1901. **Alaska:** Nagai Island, 1871–1872, *Harrington s.n.* (HT: NY!).

L. fragrans Rydb. *in* Britton, Man. fl. n. states 294. 1901. **Vermont:** Willoughby Mts., 22 Jul 1892, *Rusby s.n.* (HT: NY!; IT: NY!).

L. laxiflora BTBC **28:** 630. 1901. **Oregon:** Coast Mts., 8 Jun 1884, *Howell s.n.* (HT: NY!).

L. leptoceratitis Rydb. *in* Britton & Rydb, Bull. New York Bot. Gard. **2:** 162. 1901. **CANADA. Yukon:** Bennett City, 25 Aug 1899, *Williams s.n.* (HT: NY!).

L. leucostachys var. *robusta* BTBC **28:** 626. 1901. **Washington:** 1889, *Vasey 77* (HT: NY!).

L. media Rydb. *in* Britton, Man. fl. n. states 294. 1901. **CANADA.** Near Quebec, *Percival s.n.* (HT: NY! on 2 sheets).

L. purpurascens BTBC **28:** 269. 1901. **Colorado:** Iron Mt., 21 Jun 1900, *Rydberg & Vreeland 6414* (HT: NY!; IT: NEB).

Listera nephrophylla Mem. New York Bot. Gard. **1:** 108. 1900. **Montana:** Spanish Basin, 18 Jul 1896, *Flodman 365* (HT: NY!—designated by Rydberg in herb.).

Lysiella Mem. New York Bot. Gard. **1:** 104. 1900. T: *Orchis obtusata* Pursh.

Pipera elongata BTBC **28:** 270. 1901. **Idaho:** Priest Lake, near lower end, 26 Jul 1900, *MacDougal 168* (HT: NY!).

P. lancifolia BTBC **28:** 637. 1901. **California:** Sierra Santa Monica, 1 May 1892, *Hasse 5676* (HT: NY!).

P. leptopetala BTBC **28:** 637. 1901. **California:** Mountains east of San Diego, Jun 1850, *Parry s.n.* (HT: NY!).

P. multiflora BTBC **28:** 638. 1901. **Washington:** Gray's Harbour, 1838–1842, *Wilkes expedition 194* (HT: NY!).

OROBANCHACEAE

Myzorrhiza corymbosa BTBC **36:** 696. 1909. **Idaho:** Reynold's Creek, 2 Jul 1892, *Mulford s.n.* (HT: NY!).

OXALIDACEAE

Oxalis coloradensis BTBC **29:** 243. 1902. **Colorado:** Sangre de Cristo Creek, 2 Jul 1900, *Rydberg & Vreeland 5920* (HT: NY!; IT: NY!).

PAPAVERACEAE

Argemone rotundata BTBC **29:** 160. 1902. **NEVADA:** Diamond Mts., Jul 1868, *Watson 47* (HT: NY!).

Papaver pygmaeum BTBC **29:** 159. 1902. **MONTANA:** Mountain above Stanton Lake, 1 Aug 1894, *Williams 992* (HT: NY!; IT: MONT, US!).

PLANTAGINACEAE

Plantago myosuroides Mem. New York Bot. Gard. **1:** 369. 1900. **UTAH:** Jordan Valley, Jun 1869, *Watson 749* (HT: NY!; IT: GH!, US!).

POACEAE

Agrostis atrata BTBC **36:** 531. 1909. **CANADA. BRITISH COLUMBIA:** Yoho Valley, 6 Sep 1904, *Macoun* (herbarium *#64787*) *s.n.* (HT: NY!; IT: CAN, US!).

A. bakeri BTBC **36:** 532. 1909. **COLORADO:** Near Pagosa Peak, Aug 1899, *Baker 150* (HT: NY!; IT: US!).

Bromus flodmanii BTBC **36:** 538. 1909. **MONTANA:** Sheep Creek, 8 Aug 1896, *Flodman 187* (HT: NY!; IT: NY!).

Calamagrostis canadensis var. *acuminata* Vasey ex Rydb. & Shear, U.S.D.A. Div. Agrostol. Bull. **5:** 26. 1897. **COLORADO:** Georgetown, 17 Aug 1895, *Shear 615* (LT: US!—Hitchcock, U.S.D.A. Misc. Publ. **200:** 819. 1935).

Deschampsia alpicola BTBC **32:** 601. 1905. **COLORADO:** Pike's Peak, 4 Sep 1901, *Underwood 20* (HT: NY!).

D. pungens BTBC **39:** 103. 1912. **CANADA. ALBERTA:** Vicinity of Banff, 10 Jul 1899, *McCalla 2309* (HT: NY!; IT: CAN).

Distichlis dentata BTBC **36:** 536. 1909. **WASHINGTON:** 14 Jul 1893, *Sandberg & Leiberg 463* (HT: NY!; IT: NY!, US!).

Eatonia intermedia BTBC **32:** 602. 1905. **MONTANA:** East Gallatin Swamps, 24 Jul 1896, *Rydberg 3174* (HT: NY!; IT: US!).

E. obtusata var. *robusta* Vasey ex Rydb., Contr. U.S. Natl. Herb. **3:** 190. 1895. **NEBRASKA:** Middle Loop River near Mullen, 17 Jul 1893, *Rydberg 1807* (HT: US!; IT: NEB, NY!, US!).

Elymus marginalis BTBC **36:** 539. 1909. **CANADA. BRITISH COLUMBIA:** Lower Arrow Lake, 18 Jun 1890, *Macoun 44* (HT: NY!; IT: CAN).

E. petersonii BTBC **36:** 540. 1909. **CANADA. BRITISH COLUMBIA:** Selkirk Mts., Roger's Pass, 9 Aug 1904, *Peterson 446* (HT: NY!; IT: CAN).

E. strigosus BTBC **32:** 609. 1905. **COLORADO:** Near Boulder, 31 Jul 1886, *Letterman 553* (HT: NY!).

E. villiflorus BTBC **32:** 609. 1905. **COLORADO:** Near Boulder, Jul 1902, *Tweedy 4818* (HT: NY!; IT: US!).

E. vulpinus BTBC **36:** 540. 1909. **NEBRASKA. Grant Co.:** 3.5 miles northeast of Whitman, 29 Jul 1893, *Rydberg 1617* (HT: NY!; IT: NY!, US!).

Festuca earlei BTBC **32:** 608. 1905. **COLORADO:** La Plata Canyon, 11 Jul 1898, *Baker, Earle & Tracy 920* (HT: NY! on 2 sheets; IT: NEB, US!).

F. ingrata var. *nudata* Fl. Colorado 50. 1906. **COLORADO:** Middle Park, Aug 1892, *Beardslee s.n.* (HT: NY!; IT: NY!).

F. minutiflora BTBC **32:** 608. 1905. **COLORADO:** Cameron Pass, 13 Jul 1896, *Baker s.n.* (HT: NY!; IT: NY!).

F. saximontana BTBC **36:** 536. 1909. **CANADA. ALBERTA:** Vicinity of Banff, 28 Jul 1899, *McCalla 2331* (HT: NY!; IT: US!).

F. vallicola Mem. New York Bot. Gard. **1:** 57. 1900. **MONTANA:** Silver Bow, 8 Jul 1895, *Rydberg 2108* (HT: US!; IT: NY! 3 sheets).

Panicularia elata Nash ex Rydb., Mem. New York Bot. Gard. **1:** 54. 1900. **MONTANA:** Crazy Mts., Sweet Grass Canyon, Sep 1896, *Flodman 176* (HT: NY!).

P. nervata var. *rigida* Nash ex Rydb., Mem. New York Bot. Gard. **1:** 54. 1900. **MONTANA:** Lima, 29 Jul 1895, *Rydberg 2068* (HT: NY!).

Panicum barbipulvinatum Nash ex Rydb., Mem. New York Bot. Gard. **1:** 21. 1900. **WYOMING:** Yellowstone Park, Lower Geyser Basin, 4 Aug 1897, *Rydberg & Bessey 3544* (HT: NY! on 2 sheets).

P. capillare var. *brevifolium* Vasey ex Rydb. & Shear, U.S.D.A. Div. Agrostol. Bull. **5:** 21. 1897. **MONTANA:** Manhattan, 19 Jul 1895, *Shear 436* (HT: US!).

P. capillare var. *occidentale* Contr. U.S. Natl. Herb. **3:** 186. 1895. **NEBRASKA. Grant Co.:** 1 mile west of Whitman, 19 Sep 1893, *Rydberg 1788* (HT: US!; IT: NY! 2 sheets).

Poa callichroa BTBC **32:** 603. 1905. **COLORADO:** Dead Lake, near Pike's Peak, 14 Aug 1901, *Clements & Clements 457* (HT: NY!; IT: NEB, US!).

P. callida BTBC **36:** 533. 1909. **MONTANA:** Helena, 13 Jul 1895, *Rydberg 2145* (HT: NY!; IT: US!).

P. confusa BTBC **32:** 607. 1905. **WYOMING. Albany Co.:** Medicine Bow Mts., 28 Jul 1900, *Nelson 7787* (HT: NY!; IT: NEB, US! labeled Nash's Fork).

P. curta BTBC **36:** 534. 1909. **WYOMING:** Spread Creek, Jul 1897, *Tweedy 13* (HT: NY!).

P. helleri BTBC **36:** 534. 1909. **IDAHO. Nez Perces Co.:** Lake Waha, 20 Jun 1896, *Heller & Heller 3274* (HT: NY!; IT: NY!, US!).

P. interior BTBC **32:** 604. 1905. **WYOMING:** Headwaters of Clear Creek and Crazy Woman River, 20 Jul–15 Aug 1900, *Tweedy 3706* (HT: NY!).

P. longipila Nash ex Rydb., Mem. New York Bot. Gard. **1:** 46. 1900. **WYOMING:** Yellowstone Park, Electric Peak, 18 Aug 1897, *Rydberg & Bessey 3614* (HT: NY!; on 2 sheets; IT: MONT, NEB 2 sheets, US!).

P. macroclada BTBC **32:** 604. 1905. **COLORADO:** Roger's, 14 Aug 1901, *Baker 802* (HT: NY!; IT: US!).

P. nematophylla BTBC **32:** 606. 1905. **COLORADO. Rio Blanco Co.:** Meeker, 8 Jun 1902, *Osterhout 2601* (HT: NY!).

P. overi Brittonia **1:** 84. 1931. **SOUTH DAKOTA. Custer Co.:** Gray Rocks, 19 Jul 1929, *Over 18100* (HT: SDU—fide van Bruggen in litt. 1984; IT: NY!).

P. phoenicea BTBC **32:** 605. 1905. **COLORADO:** Peak Valley, 21 Aug 1901, *Clements & Clements 466* (HT: NY!; IT: NEB, US!).

P. plattensis Brittonia **1:** 84. 1931. **NEBRASKA. Banner Co.:** Lawrence Fork, 8 Jul 1891, *Rydberg 461* (HT: NY!).

P. pseudopratensis Scribner & Rydb. *in* Rydb., Contr. U.S. Natl. Herb. **3:** 531. 1896. **SOUTH DAKOTA:** Hot Springs, 13 Jun 1892, *Rydberg 1151* (HT: US!; IT: NY!).

P. pudica BTBC **32:** 603. 1905. **COLORADO:** Steven's Mine, 23 Aug 1895, *Rydberg 2443* (HT: NY!; IT: NEB, NY!, US!).

P. scaberrima BTBC **36:** 534. 1909. **IDAHO:** Beaver Canyon, 27 Jun 1895, *Rydberg 2055* (HT: NY!; IT: NY!, US!).

P. subreflexa BTBC **36:** 535. 1909. **COLORADO:** Mountains north of Steamboat Springs, 26 Jul 1891, *collector unknown* (from Herb. State Agric. College Colo.) *3731* (HT: NY!).

P. subtrivialis BTBC **36:** 535. 1909 **WYOMING. Sheridan Co.:** Big Horn Mts., Jul 1899, *Tweedy 2141* (HT: NY!).

P. tricholepis BTBC **32:** 606. 1905. **COLORADO:** Near Pagosa Peak, Aug 1899, *Baker 210* (HT: NY!; IT: US!).

P. truncata BTBC **32:** 607. 1905. **COLORADO. Summit Co.:** Dillon, 26 Aug 1896, *Clements 373* (HT: NY!).

Sporobolus aristatus BTBC **28:** 266. 1901. **WYOMING: Sheridan Co.:** Big Horn Mts., 1899, *Tweedy 2196* (HT: NY!).

Trisetum majus Fl. Colorado 33, 34. 1906. **COLORADO. [Park Co.]:** Pen[nsylvania] Gulch, 1884, *Vasey s.n.* (HT: NY!; IT: US!).

POLEMONIACEAE

Gilia candida BTBC **28:** 29. 1901. **COLORADO:** Near La Veta, 11 Jul 1900, *Vreeland 602* (HT: NY!; IT: RM).

G. cephaloidea BTBC **24:** 293. 1897. **MONTANA:** Lima, 29 Jun 1895, *Rydberg 2764* (HT: NY!).

G. crandallii BTBC **31:** 634. 1904 [1905]. **COLORADO:** Durango, 28 Jun 1898, *Crandall 2053* (HT: NY!; CT: UC!).

G. frutescens BTBC **40:** 471. 1913. **UTAH:** Springdale, 14 May 1894, *Jones 5247* (HT: NY!; IT: US! dated 16 May).

G. parvula Mem. New York Bot. Gard. **1:** 320. 1900. **WYOMING:** Yellowstone Park, Swan Lake, Jun 1885, *Tweedy 823* (HT: YU!; IT: NY!).

G. polyantha BTBC **31:** 634. 1904 [1905]. **COLORADO:** Pagosa Springs, 28 Jul 1899, *Baker 538* (HT: NY!; IT: US!).

G. roseata BTBC **31:** 633. 1904 [1905]. **COLORADO:** Grand Junction, May 1892, *Eastwood s.n.* (HT: NY!; IT: DS!, US!).

G. scariosa BTBC **31:** 632. 1904 [1905]. **COLORADO:** Veta Pass, 15 Jul 1896, *Shear 3590* (HT: NY!).

G. spergulifolia BTBC **31:** 633. 1904 [1905]. **WYOMING:** Big Horn Mts., headwaters of Tongue River, Jul 1898, *Tweedy 576* (HT: NY!).

G. straminea BTBC **40:** 472. 1913. **UTAH:** St. George, 1877, *Palmer 325* in part (HT: NY!).

G. subacaulis BTBC **30:** 261. 1903. **WYOMING:** Point of Rocks, 1901, *Merrill & Wilcox 607* (HT: NY!).

G. tenuituba BTBC **40:** 472. 1913. **UTAH:** Beaver City, 1877, *Palmer 329* (HT: NY!; IT: US! no data).

G. tridactyla Fl. Rocky Mts. 692, 1065. 1917 [1918]. **UTAH:** Brigham Peak, Marysvale, 29 Aug 1894, *Jones 5949* (HT: NY!; IT: US!).

G. tweedyi BTBC **31:** 634. 1904 [1905]. **WYOMING. Carbon Co.:** Encampment, 20 Jun 1901, *Tweedy 4422* (HT: NY!).

Leptodactylon brevifolium BTBC **40:** 474. 1913. **UTAH:** Juniper Range, May–Oct 1898, *Purpus 6306* (HT: US!; CT: NY!).

Phlox collina Mem. New York Bot. Gard. **1:** 316. 1900. **MONTANA. Madison Co.:** May 1888, *Tweedy 154* (HT: YU!; IT: NY!).

P. costata Mem. New York Bot. Gard. **1:** 315. 1900. **MONTANA:** Cedar Mts., 16 Jul 1897, *Rydberg & Bessey 4807* (HT: NY!; IT: MONT, NEB, US!).

P. diapensioides Mem. New York Bot. Gard. **1:** 317. 1900. **MONTANA. Madison Co.:** 12 Oct 1887, *Tweedy 282* (HT: YU!—designated by Rydberg in herb.; IT: NY!).

P. scleranthifolia Mem. New York Bot. Gard. **1:** 313. 1900. **SOUTH DAKOTA:** Black Hills, Hot Springs, 13 Jun 1892, *Rydberg 880* (HT: NY!; IT: NEB).

Polemonium columbianum BTBC **40:** 477. 1913. **IDAHO:** Coeur d'Alene Mts., divide between St. Joseph and Clearwater Rivers, 9 Jul 1896, *Leiberg 1205* (HT: NY!; IT: US!).

P. delicatum BTBC **28:** 29. 1901. **COLORADO:** West Spanish Peak, 9 Jul 1900, *Rydberg & Vreeland 5720* (HT: NY!; IT: NEB).

P. grayanum BTBC **31:** 635. 1904 [1905]. **COLORADO:** Gray's Peak, 23 Aug 1895, *Rydberg s.n.* (HT: NY!; IT: NY!).

P. robustum BTBC **31:** 635. 1904 [1905]. **COLORADO:** La Plata Mts., Bob Creek, 1 Jul 1898, *Baker, Earle & Tracy 274* (HT: NY!; IT: NY!, US!).

P. scopulinum E. L. Greene ex Rydb., Fl. Colorado 280. 1906. **COLORADO:** Near Pagosa Peak, 6 Aug 1899, *Baker 546* (HT: NY!; IT: US!).

P. speciosum BTBC **28:** 29. 1901. **COLORADO:** Mt. Garfield, 1900, *Clements s.n.* (HT: NY!).

Tintinabulum Fl. Rocky Mts. 698, 1065. 1917 [1918]. T: *Gilia filiformis* Parry.

POLYGONACEAE

Chorizanthe spathulata Small ex Rydb., BTBC **39:** 309. 1912. **IDAHO:** Big Butte Station, 23 Jun 1893, *Palmer 230* (HT: NY!).

Eriogonum biumbellatum BTBC **39:** 306. 1912. **UTAH:** Fish Lake around Twin Creeks, 8 Aug 1905, *Rydberg & Carlton 7609* (HT: NY!—designated by Rydberg in herb.; IT: RM, US!).

E. brandegei Fl. Rocky Mts. 220, 1061. 1917 [1918]. **COLORADO:** Southern Colorado, 1 Sep 1873, *Brandegee s.n.* (HT: NY!).

E. chrysops Fl. Rocky Mts. 220, 1061. 1917 [1918].

OREGON: Northern Steen's Mts., 10 Jun 1901, *Cusick 2558* (HT: NY!; IT: US!).

E. clutei Amer. Bot. **27:** 61. 1921. **ARIZONA:** Cameron, near Tuba, 15–31 Jul 1920, *Clute 71a* (HT: NY!).

E. contortum Small ex Rydb., Fl. Colorado 101–104, 107. 1906. **COLORADO:** Grand Junction, May 1892, *Eastwood s.n.* (HT: NY!; IT: CAS!, NY!).

E. idahoense BTBC **39:** 307. 1912. **IDAHO. Washington Co.:** Weiser, 7 Jul 1899, *Jones 6511* (HT: NY!; IT: US! 3 sheets).

E. lagopus Fl. Rocky Mts. 222, 1061. 1917 [1918]. **WYOMING. Sheridan Co.:** Dayton, Sep 1899, *Tweedy 2624* (HT: NY!).

E. latum Small ex Rydb., Mem. New York Bot. Gard. **1:** 121. 1900. **MONTANA:** Cottonwood Creek, 30 Jul 1896, *Flodman s.n.* (HT: NY!; IT: CAN, NY! numbered *376*).

E. mancum Fl. Rocky Mts. 220, 1061. 1917 [1918]. **MONTANA. Beaver Head Co.:** Jul 1888, *Tweedy 106* (HT: NY!; IT: MONT).

E. medium Fl. Rocky Mts. 220, 1061. 1917 [1918]. **UTAH:** Mt. Nebo, 15 Aug 1905, *Rydberg & Carlton 7594* (HT: NY!; IT: CAN, NY!, RM, US!).

E. nebraskense Fl. Rocky Mts. 224, 1061. 1917 [1918]. **NEBRASKA. Kimball Co.:** 12 Jul 1891, *Rydberg 337* (HT: NY!; IT: NY!, US! both dated 12 Aug).

E. ochroleucum Small ex Rydb., Mem. New York Bot. Gard. **1:** 123. 1900. **MONTANA. Gallatin Co.:** Spanish Basin, 23 Jun 1897, *Rydberg & Bessey 5343* (HT: NY!; IT: CAN, MONT, NEB, NY!).

E. polyphyllum Small ex Rydb., Mem. New York Bot. Gard. **1:** 119. 1900. **MONTANA:** Old Hollowtop near Pony, 9 Jul 1897, *Rydberg & Bessey 5325* (HT: NY! on 2 sheets; IT: CAN, MONT, NEB).

E. spathuliforme BTBC **39:** 307. 1912. **UTAH:** Near Belknap, 12 Jun 1900, *Stokes s.n.* (HT: NY!; IT: US!).

Persicaria iowensis Brittonia **1:** 87. 1931. **IOWA:** Ames, 29 Jul 1902, *Buchanan s.n.* (HT: ISC—fide Lammers in litt. 1985; IT: NY!).

Polygonum paronychioides Small ex Rydb., Mem. New York Bot. Gard. **1:** 130. 1900. **MONTANA:** South of Birch Lake, 10 Aug 1883, *Canby 279* (HT: NY!; IT: NY! 2 sheets).

P. unifolium Small ex Rydb., Mem. New York Bot. Gard. **1:** 129. 1900. **MONTANA:** Cedar Mts., 16 Jul 1897, *Rydberg & Bessey 5367* (HT: NY!—designated by Rydberg in herb.; IT: MONT).

Rumex praecox BTBC **33:** 137. 1906. **COLORADO:** Grayback mining camps and Placer Gulch, 25–27 Jun 1900, *Rydberg & Vreeland 6328* (HT: NY!).

POLYPODIACEAE

Pellaea pumila Mem. New York Bot. Gard. **1:** 4. 1900. **SOUTH DAKOTA:** Black Hills, Bull Springs west of Custer, 27 Jul 1892, *Rydberg 1191* (HT: NY!).

Thelypteris oreopteris var. *hesperia* Slosson ex Rydb., Fl. Rocky Mts. 1044, 1069. 1917 [1918]. **ALASKA:** Attis Island, 29 Aug 1891, *Macoun 153* (HT: NY!).

PORTULACACEAE

Calandrinia arizonica NAF **21:** 292. 1932. **ARIZONA:** Wet sands of the Rillita, 13 Apr 1881, *Pringle s.n.* (HT: GH!).

C. filifolia NAF **21:** 293. 1932. **MEXICO. BAJA CALIFORNIA:** San Quentin Bay, Jan 1889, *Palmer 733* (HT: NY!).

C. heterophylla NAF **21:** 293. 1932. **CALIFORNIA. San Diego Co.:** Old Mission Dam, 10 Apr 1904, *Chandler 5152* (HT: NY!).

C. muricata NAF **21:** 293. 1932. **CALIFORNIA:** San Diego, Apr 1906, *Brandegee s.n.* (HT: NY!).

C. stenophylla NAF **21:** 292. 1932. **CALIFORNIA:** Alameda, 17 Jun 1891, *Greene s.n.* (HT: NY!).

C. tenella NAF **21:** 293. 1932. **OREGON:** Sauvie's Island, 20 Apr 1885, *Howell s.n.* (HT: NY! on 2 sheets; IT: US!).

Calyptridium pygmaeum Parish ex Rydb., NAF **21:** 320. 1932. **CALIFORNIA:** San Bernardino Mts., Bear Valley, Jun 1886 [1885 in publication], *Parish 1803* (HT: DS!; IT: US!, YU!).

Claytonia bellidifolia NAF **21:** 301. 1932. **OREGON:** Highest mountains in eastern Oregon, 1897, *Cusick 1777* (HT: US!; CT: NY!).

C. obovata NAF **21:** 299. 1932. **OREGON:** Abbott's Butte, 6 Jul 1899, *Leiberg 4262* (HT: US!).

C. rosea BTBC **31:** 404. 1904. **COLORADO:** La Veta, 18 May 1900, *Rydberg & Vreeland 6300* (HT: NY!; IT: NY!).

Crunocallis BTBC **33:** 139. 1906. T: *Claytonia chamissoi* Ledeb.

Erocallis BTBC **33:** 139. 1906. T: *Claytonia triphylla* S. Wats.

Lewisia minor NAF **21:** 327. 1932. **CALIFORNIA. Ventura Co.:** Griffins, Jul 1902, *Elmer 3886* (HT: NY!; IT: CAS!, US!).

Limnalsine NAF **21:** 295. 1932. T: *Claytonia diffusa* Nutt.

Limnia bracteosa NAF **21:** 305. 1932. **CALIFORNIA. Marin Co.:** 3½ miles north of Point Reyes Light House, 10 Mar 1928, *Bacigalupi 1818* (HT: NY!; IT: DS!).

L. diaboli NAF **21:** 312. 1932. **CALIFORNIA:** Mt. Diablo, 28 Apr 1868, *Kellogg & Harford 100* (HT: NY!; IT: US! no data).

L. guadalupensis NAF **21:** 311. 1932. **MEXICO. BAJA CALIFORNIA:** Guadalupe Island, 1875, *Palmer 15* in part (HT: NY!).

L. hassei NAF **21:** 311. 1932. **CALIFORNIA. Los Angeles Co.:** 25 Apr 1888, *Hasse 4120* (HT: NY!).

L. membranacea NAF **21:** 308. 1932. **UTAH. Cache Co.:** Wellsville Canyon, 8 Mar 1911 [1910 in publication], *Zundel 176* (HT: NY!; IT: DS).

L. mexicana NAF **21:** 309. 1932. **MEXICO.** Nevada de Toluca, 15 Oct 1903, *Rose & Painter 7924* (HT: US not found).

L. platyphylla NAF **21:** 307. 1932. **CALIFORNIA:** Dos Pueblos near Gaviota Pass, 2–5 Apr 1861, *Brewer 384* (HT: US!).

L. utahensis BTBC **39:** 314. 1912. **UTAH:** St. George, 1877, *Palmer 56* (HT: NY!).

Montia andina NAF **21:** 316. 1932. **MEXICO.** Ixtaccihuatl, Oct 1905, *Purpus 1785* (HT: NY!; IT: US!).

M. funstonii NAF **21:** 316. 1932. **CALIFORNIA. Tulare Co.:** Whitney Meadows, 19 Aug 1891, *Coville & Funston 1619* (HT: US!; CT: DS!, NY!).

M. minima NAF **21:** 315. 1932. **MEXICO. LOWER CALIFORNIA:** San Pedro Martín, 18 & 22 May 1893, *Brandegee s.n.* (HT: UC! both mounted on one sheet; CT: NY!).

M. stenophylla NAF **21:** 315. 1932. **MEXICO. LOWER CALIFORNIA:** Valley of Palms, 15 Apr 1882, *Jones s.n.* (HT: NY!).

Oreobroma congdonii NAF **21:** 322. 1932. **CALIFORNIA. Mariposa Co.:** Hennesey Ranch, 31 May 1883, *Congdon s.n.* (HT: DS! no date).

O. glandulosum NAF **21:** 325. 1932. **CALIFORNIA:** Mt. Dana, Jul 1902, *H. M. Hall & Babcock 3611* (HT: NY!; IT: US!).

O. mexicanum NAF **21:** 326. 1932. **MEXICO. [VERACRUZ]:** Mt. Orizaba, 25–26 Jul 1901, *Rose & Hay 5769* (HT: US!).

Spraguea caespitosa NAF **21:** 318. 1932. **CALIFORNIA. Los Angeles Co.:** San Gabriel Mts., North Baldy Mt., 3 Jul 1908, *Abrams & McGregor 607* (HT: NY!; IT: US!).

S. hallii NAF **21:** 318. 1932. **CALIFORNIA:** San Jacinto Mts., 22 Jul 1897, *H. M. Hall 714* (HT: UC!).

S. irregularis NAF **21:** 318. 1932. **CALIFORNIA:** San Bernadino Mts., Bear Valley, 3 Jun 1892, *Parish 2339* (HT: NY!).

PRIMULACEAE

Androsace albertina BTBC **40:** 462. 1913. **CANADA. ALBERTA:** Lake Agnes National Park, Banff, Aug 1897, *Van Brunt & Van Brunt 77* (HT: NY!).

A. puberulenta BTBC **30:** 260. 1903. **COLORADO:** Near Veta Pass, 20 Jun 1900, *Rydberg & Vreeland 5772* (HT: NY!).

A. simplex BTBC **40:** 462. 1913. **MONTANA:** Missoula, May 1897, *Elrod 33* (HT: NY!).

Dodecatheon acuminatum Mem. New York Bot. Gard. **1:** 304. 1900. **MONTANA:** Missouri River above mouth of Sand Coulee, 30 Apr 1885, *Anderson s.n.* (HT: NY!; IT: US!).

D. cylindrocarpum Mem. New York Bot. Gard. **1:** 305. 1900. **MONTANA:** Bridger Mts., 17 Jun 1897, *Rydberg & Bessey 4674* (LT: NY!—Thompson, Contr. Dudley Herb. **4:** 136. 1953; ILT: NEB, US!).

D. multiflorum BTBC **31:** 631. 1904 [1905]. **COLORADO:** Sangre de Cristo Creek, 2 Jul 1900, *Rydberg & Vreeland 5781* (HT: NY!—designated by Rydberg in herb.; IT: NY!).

D. pubescens Mem. New York Bot. Gard. **1:** 306. 1900. **MONTANA:** Missoula, Apr 1883, *Tweedy s.n.* (HT: YU!; CT: NY!).

D. pulchrum Mem. New York Bot. Gard. **1:** 305. 1900. **WYOMING:** Yellowstone Park, Indian Creek, Jun 1885, *Tweedy 432* in part (HT: YU!; IT: US!; CT: NY!).

D. radicatum var. *sinuatum* BTBC **31:** 631. 1904 [1905]. **COLORADO. Larimer Co.:** 14 May 1890, *Crandall s.n.* (HT: NY!).

D. uniflorum Mem. New York Bot. Gard. **1:** 307. 1900. **MONTANA:** Old Hollowtop, near Pony, 7 Jul 1897, *Rydberg & Bessey 4668* (LT: NY!—Thompson, Contr. Dudley Herb. **4:** 147. 1953).

Primula americana BTBC **28:** 500. 1901. **MONTANA:** Deer Lodge, 9 July 1895, *Rydberg 2746* (HT: NY!).

P. parryi var. *brachyantha* Mem. New York Bot. Gard. **1:** 302. 1900. **MONTANA. Park Co.:** Sheep Mt., 17 Jul 1887, *Tweedy 82* (HT: YU!; IT: MONT).

P. specuicola BTBC **40:** 461. 1913. **UTAH:** San Juan River near bluffs, 25–29 Aug 1911, *Rydberg 9882* (HT: NY!; IT: US!, UT!).

PYROLACEAE

Chimaphila acuta NAF **29:** 31. 1914. **ARIZONA:** Tonto Basin, 2 Aug 1887, *Mearns 136* (HT: NY!; IT: NY!).

C. dasystemma Torrey ex Rydb., NAF **29:** 32. 1914. **MEXICO.** *Coulter s.n.* (HT: NY!).

C. guatemalensis NAF **29:** 32. 1914. **GUATEMALA. ZACATEPÉQUEZ:** Santiago, 1891, *Gómez 795* (HT: NY!; IT: US!).

C. occidentalis NAF **29:** 30. 1914. **IDAHO. Latah Co.:** Pine Creek near Farmington, 28 Jun 1892, *Sandberg, MacDougal & Heller 519* (HT: NY! on 2 sheets; IT: CAS!).

Pyrola borealis NAF **29:** 24. 1914. **CANADA. [NORTHWEST TERRITORIES]:** Mackenzie River, 1861–1862, *Onion, Kennicott & Hardisty s.n.* (HT: NY!; IT: NY!).

P. gormanii NAF **29:** 25. 1914. **CANADA. YUKON TERRITORY:** Dry Gulch, 14 Jun 1899, *Gorman 1029* in part (HT: NY!; IT: CAN).

RANUNCULACEAE

Aconitum atrocyaneum BTBC **29:** 150. 1902. **COLORADO:** Boreas, 24 July 1897, *Crandall s.n.* (HT: NY!; IT: NY!, US! numbered *311*).

A. divaricatum Fl. Rocky Mts. 314, 1062. 1917 [1918]. **UTAH:** City Creek Canyon, 9 Aug 1884, *Leonard 204/182* (HT: NY! on 2 sheets).

A. glaberrimum BTBC **29:** 151. 1902. **SOUTHERN UTAH, NORTHERN ARIZONA:** 1877, *Palmer 11* (HT: NY!; IT: NY!).

A. insigne E. L. Greene ex Rydb., Fl. Colorado 139. 1906. **COLORADO:** Near Pagosa Peak, Aug 1899, *Baker 317* (HT: NY!; IT: US! 2 sheets).

A. patens Fl. Rocky Mts. 315, 1062. 1917 [1918]. **IDAHO. Latah Co.:** Viola, 14 Jul 1896, *Elmer 13* (HT: NY!; IT: US! dated just 1896).

A. porrectum BTBC **29:** 150. 1902. **COLORADO:** Coffee Pot Spring, 10 Aug 1895, *Crandall s.n.* (HT: NY!).

A. tenue BTBC **29:** 149. 1902. **SOUTH DAKOTA:** Black Hills, Sylvan Lake, 18 Jul 1892, *Rydberg 507* (HT: NY!; IT: NEB 2 sheets, US!).

A. vaccarum Torreya **26:** 31. 1926. **WEST VIRGINIA. Pendleton Co.:** Alleghany Mts., Spruce Knob, 21–22 Jun 1925, *Rydberg 9206* (HT: NY!; IT: NY!, US!).

Actaea eburnea Mem. New York Bot. Gard. **1:** 153. 1900. **MONTANA:** Bridger Mts., 18 Jun 1897, *Rydberg & Bessey 4062* (HT: NY!—designated by Rydberg in herb.; IT: MONT).

Anemone lithophila BTBC **29:** 152. 1902. **MONTANA:** Little Belt Mts., nine miles from Barker, 18 Aug 1896, *Flodman 459* (HT: NY!; IT: NY!).

A. piperi Britton ex Rydb., BTBC **29:** 153. 1902. **IDAHO. Latah Co.:** 16 Jun 1893, *Piper 1469* (HT: NY!).

A. tuberosa BTBC **29:** 151. 1902. **ARIZONA:** Sierra Tucson, 8 Mar 1884, *Pringle s.n.* (HT: NY!; IT: DS!).

Aquilegia columbiana BTBC **29:** 145. 1902. **WASHINGTON. Chehalis Co.:** Montesano, 13 Jun 1898, *Heller & Heller 3936* (HT: NY!).

A. saximontana Rydb. ex B.L. Robinson *in* A. Gray, Syn. Fl. N. Amer. I, **1:** 43. 1895. **COLORADO.** Gray's Peak, 23 Aug 1895, *Rydberg s.n.* (LT: NY!—Munz, Gentes Herb. **7:** 28. 1946).

A. thalictrifolia BTBC **29:** 145. 1902. **COLORADO:** Canyon City, 23 Jun 1873, *Greene s.n.* (HT: NY!; IT: CAS!).

Atragene grosseserrata BTBC **29:** 156. 1902. **IDAHO. Palouse Co.:** About Lake Coeur d'Alene, Jun–Jul 1892, *Aiton s.n.* (HT: NY!).

A. pseudo-alpina var. *diversiloba* BTBC **31:** 409. 1904. **COLORADO:** Mountains near Denver, Aug 1869, *B.H. Smith s.n.* (HT: NY!).

Caltha uniflora Mem. New York Bot. Gard. **1:** 474. 1900. **MONTANA:** Haystack Peak, Aug 1899, *Koch s.n.* (HT: NY!; IT: MONT).

Clematis eriophora BTBC **29:** 154. 1902. **COLORADO:** Vicinity of Horsetooth, 26 Apr 1896, *Cowan s.n.* (HT: NY!; IT: CS).

C. missouriensis Rydb. *in* Britton, Man. fl. n. states 421. 1901. **NEBRASKA:** Lincoln, 1886, *Webber s.n.* (HT: NY!; IT: NY!).

Delphinium albescens BTBC **26:** 583. 1899. **NEBRASKA:** Lincoln, Jun 1887, *Webber s.n.* (HT: NY!).

D. alpestre BTBC **29:** 146. 1902. **COLORADO:** Mountains northwest of Como, 31 Jul 1895, *Crandall & Cowan 1948* [*1848* in publication] (HT: CS—fide Wilken in litt. 1984).

D. bicolor var. *montanense* Mem. New York Bot. Gard. **1:** 157. 1900. **MONTANA:** Deer Lodge, Jun 1888, *Traphagen s.n.* (HT: NY!—designated by Rydberg in herb.).

D. brownii BTBC **29:** 148. 1902. **CANADA. ALBERTA:** Banff, 13 Aug 1893, *Brown s.n.* (HT: NY!).

D. canmorense Fl. Rocky Mts. 312, 1061. 1917 [1918]. **CANADA. BRITISH COLUMBIA:** Canmore, 1 Jul 1885, *Macoun s.n.* (HT: NY!).

D. coelestinum BTBC **39:** 320. 1912. **SOUTHERN UTAH, NORTHERN ARIZONA:** 1877, *Palmer 11* (HT: NY!; IT: CAN, US! numbered *10*).

D. diversicolor BTBC **29:** 149. 1902. **MONTANA. Beaver Head Co.:** Rattlesnake Creek, Jun 1888 [1887 in publication], *Tweedy 34* (HT: NY!; IT: YU).

D. elongatum BTBC **29:** 148. 1902. **COLORADO. Larimer Co.:** 19 Aug 1895, *Crandall & Cowan 15* (HT: NY!).

D. geraniifolium BTBC **26:** 583. 1899. **ARIZONA:** Charles' Valley, 1 Aug 1883, *Rusby s.n.* (HT: NY!; IT: US!).

D. glaucescens Mem. New York Bot. Gard. **1:** 155. 1900. **MONTANA:** Yellowstone Park, Electric Peak, 18 Aug 1897, *Rydberg & Bessey 4078* (HT: NY!; IT: MONT, NEB 2 sheets, US!).

D. glaucescens var. *multicaule* Mem. New York Bot. Gard. **1:** 156. 1900. **MONTANA:** Cedar Mt., 16 Jul 1897, *Rydberg & Bessey 4071* (HT: NY!; IT: MONT, NEB, US!).

D. helleri BTBC **39:** 321. 1912. **IDAHO. Nez Perces Co.:** Lewiston, 24 Apr 1896, *Heller & Heller 2951* (HT: NY!; IT: NY!, US!).

D. leonardii BTBC **39:** 320. 1912. **UTAH:** Garfield, 30 May 1884, *Leonard 205* (HT: NY! on 2 sheets).

D. macroseratilis BTBC **26:** 585. 1899. **TEXAS. Tom Greene Co.:** 1879, *Tweedy s.n.* (HT: NY!).

D. multiflorum BTBC **29:** 147. 1902. **WYOMING:** Spread Creek, Jul 1897, *Tweedy 179* (HT: NY!).

D. ramosum BTBC **28:** 276. 1901. **COLORADO:** North Cheyenne Canyon, 14 Jul 1896, *Bessey s.n.* (HT: NY!).

D. robustum BTBC **28:** 276. 1901. **COLORADO:** Wahatoya Creek, 7 Jul 1900, *Rydberg & Vreeland 6217* (HT: NY! on 2 sheets).

D. viscidum BTBC **39:** 322. 1912. **WYOMING:** Near Tie Siding, 16 Jul 1896, *Osterhout s.n.* (HT: NY!).

D. wootoni BTBC **26:** 587. 1899. **NEW MEXICO:** Organ Mts., 20 Apr 1893, *Wooton s.n.* (HT: NY!; IT: CAS! dated 30 Apr and numbered *272*, US! dated 30 Apr).

D. xylorrhizum BTBC **39:** 321. 1912. **MONTANA:** Lima, 1 Jul 1895, *Shear 3429* (HT: NY!).

Ranunculus helleri BTBC **29:** 158. 1902. **IDAHO. Kootenai Co.:** near Lake Pend d'Oreille, 5 Aug 1892, *Sandberg, MacDougal & Heller 842* (HT: NY!; IT: CAS!).

R. montanensis Mem. New York Bot. Gard. **1:** 166. 1900. **MONTANA:** Helena, Jun 1891, *Kelsey s.n.* (HT: NY!).

R. rivularis BTBC **39:** 319. 1912. **NEVADA:** Huntington Valley, Aug 1868, *Watson 27* (HT: NY!).

R. saxicola Mem. New York Bot. Gard. **1:** 164. 1900. **MONTANA:** Cedar Mt., 16 Jul 1897, *Rydberg & Bessey 4112* (LT: NY—Benson, BTBC **68:** 654. 1940, not found; ILT: CAN, MONT, NEB, POM, US!).

R. stenolobus BTBC **29:** 159. 1902. **WYOMING:** Headwaters of Cliff Creek, 9–18 Aug 1900, *Curtis s.n.* (HT: NY!; IT: NY!).

R. utahensis BTBC **29:** 158. 1902. **UTAH:** Wasatch Mts., Alta, 31 Jul 1879, *Jones 1130* (HT: NY!; IT: NY!, POM, UTC!).

Thalictrum columbianum BTBC **39:** 320. 1912. **WASHINGTON. Okanogan Co.:** Loomiston, Aug 1897, *Elmer 599* (HT: NY!; IT: US!).

T. hypoglaucum Brittonia **1:** 88. 1931. **TEXAS:** Dallas, 3 Jul 1872, *E. Hall 2* (HT: NY!; IT: US!).

T. megacarpum Torrey ex Rydb., Fl. Rocky Mts. 290. 1917 [1918]. **[WYOMING]:** 26–31 Aug 1842, *Fremont s.n.* (HT: NY!).

T. stipitatum Fl. Rocky Mts. 290. 1917 [1918]. **COLORADO:** Minnehaha, 1 Jul 1901, *Clements & Clements 243* (HT: NY!; IT: US!).

RHAMNACEAE

Ceanothus subsericeus BTBC **31:** 564. 1904. **COLORADO. Larimer Co.:** 20 Jul 1895, *Cowen s.n.* (HT: NY!).

ROSACEAE

Adenostoma brevifolium Nutt. ex Rydb., NAF **22:** 396. 1913. CALIFORNIA. San Diego, *Nuttall s.n.* (ST: NY!); San Diego, 1875, *Palmer 79* (ST: NY!).

Agrimonia pringlei NAF **22:** 395. 1913. **MEXICO.** VERACRUZ: Near Jalapa, 30 Nov 1903, *Pringle 11876* (HT: GH!; CT: NY! 2 sheets).

Amelanchier carrii Brittonia **1:** 89. 1931. SOUTH DAKOTA. Deadwood, 18 Jul 1913, *Carr 75* (HT: NY!).

Aphanes australis NAF **22:** 380. 1908. Near Washington, D.C., 10 Jun 1898. *Steele s.n.* (HT: NY!).

A. macrosepala NAF **22:** 380. 1908. OREGON. **Jackson Co.:** Wimer, 17 May 1892, *Hammond 116* (HT: NY!).

Argentina argentea BTBC **33:** 143. 1906. **CANADA.** Assiniboia, Wood Mt., 4 Jun 1895, *Macoun* [herbarium *#10471*] *s.n.* (LT: NY!—Rydberg, NAF **22:** 353. 1908).

A. babcockiana NAF **22:** 354. 1908. NEW YORK: Westminster Park, 1900, *Babcock s.n.* (HT: NY!).

A. litoralis NAF **22:** 354. 1908. **CANADA.** NOVA SCOTIA: Yarmouth, 22–29 Jun 1901, *Howe & Lang 1* (HT: NY!; IT: GH!).

A. occidentalis NAF **22:** 354. 1908. CALIFORNIA. **Solano Co.:** Suisun, 6 Jun 1903, *Baker 3217* (HT: NY!; IT: CAS!).

A. subarctica NAF **22:** 354. 1908. ALASKA: Yukon River near Palisades, 14 Jul 1903, *Hollick s.n.* (HT; NY!).

Aruncus allegheniensis NAF **22:** 256. 1908. MARYLAND: Baltimore, 1867, *LeRoy s.n.* (HT: NY!).

A. pubescens NAF **22:** 256. 1908. ILLINOIS: Peoria, Jul 1903, *McDonald s.n.* (HT: NY!; IT: NY!).

Cercocarpus alnifolius NAF **22:** 421. 1913. CALIFORNIA: Santa Catalina Island, Avalon, Jun 1896, *Trask s.n.* (HT: US!; CT: NY!).

C. argenteus NAF **22:** 422. 1913. TEXAS. **Randall Co.:** Rocky Bluffs on the Red River, 9 Jun 1901, *Eggert s.n.* (HT: MO—fide Moran in litt. 1986; CT: NY!).

C. douglasii NAF **22:** 421. 1913. CALIFORNA: *Douglas s.n.* (HT: NY!).

C. flabellifolius NAF **22:** 422. 1913. UTAH: Near Glenwood, 29 May 1875, *Ward 122* (HT: NY!; IT: US!).

C. hypoleucus NAF **22:** 424. 1913. MONTANA: Melrose, 7 Jul 1895, *Rydberg 2695* (HT: NY!; IT: NY!).

C. macrourus NAF **22:** 420. 1913. CALIFORNIA. **Modoc Co.:** 20 Jun & 23 Jul 1893, *M. Baker s.n.* (HT: NY!; IT: NY! 2 sheets).

C. rotundifolius NAF **22:** 421. 1913. CALIFORNIA. **Los Angeles Co.:** Mar 1901, *Grant 3488* (HT: NY!).

Chamaebatiaria glutinosa NAF **22:** 258. 1908. NEVADA. **Nye Co.:** Mammoth Range, a few miles from Ellsworth, 14 Sep 1868, *Howard s.n.* (HT: NY!).

Comarella Mem. Dept. Bot. Columbia Coll. **2:** 156. 1898. T: *Horkelia multifoliata* Torrey.

Cowania davidsonii NAF **22:** 416. 1913. ARIZONA: Blue River, 8 Sep 1902. *Davidson 754* (HT: NY!; IT: DS!).

Drymocallis albida NAF **22:** 375. 1908. WASHINGTON. **Klickitat Co.:** Bingen, Jun & Jul 1893, *Suksdorf 2209* (HT: NY!; IT: GH!, US!; CT: DS!, UC!).

D. amplifolia NAF **22:** 372. 1908. WASHINGTON. **Klickitat Co.:** Near the Columbia River, Jun 1893, 1894, *Suksdorf 1761* (HT: NY!; IT: NY!, UC!, US!).

D. arizonica NAF **22:** 373. 1908. ARIZONA: Mormon Lake, 7 Jun 1898, *MacDougal 64* (HT: NY! on 2 sheets; IT: GH!, US!).

D. corymbosa NAF **22:** 369. 1908. MONTANA. **Gallatin Co.:** Spanish Basin, 24 Jun 1897, *Rydberg & Bessey 4348* (HT: NY!; IT: NEB, US!).

D. cuneifolia Mem. Dept. Bot. Columbia Coll. **2:** 204. 1898. CALIFORNIA: San Bernardino Mts., near Green lead mines, Jun 1886, *Parish 1818* (HT: NY!; IT: GH!; CT: DS!).

D. foliosa NAF **22:** 371. 1908. MONTANA: Bridger Mts., 28 Jul 1896, *Flodman 596* (HT: NY!; IT: US!).

D. glabrata Mem. Dept. Bot. Columbia Coll. **2:** 201. 1898. WASHINGTON. **Kittitas Co.:** 8 miles southwest of Ellensburg, Jun 1897, *Elmer 412* (HT: NY! on 2 sheets; IT: BKL, US!).

D. glandulosa var. *monticola* Mem. Dept. Bot. Columbia Coll. **2:** 199. 1898. CALIFORNIA: Head of Tuolumne, 1863, *Brewer 1767* (ST: GH!); Soda Springs, 1863, *Brewer 1705* (ST: GH!); [no data,] *Brewer 2715* (ST: GH!); *Wright 15*; **Amador Co.:** New York Falls, 12 Jul 1892, *Hansen 296* (ST: DS!). IDAHO: 1896, *Henderson 3599.* MONTANA: 1891, *Kelsey s.n.* NEVADA: 1864, *Anderson s.n.* OREGON: *Cronkite s.n.*

D. glutinosa var. *neo-mexicana* Mem. Dept. Bot. Columbia Coll. **2:** 197. 1898. NEW MEXICO: 1847, *Fendler 179;* 1867, *Bell s.n.*; *Newberry s.n.*

D. gracilis BTBC **28:** 177. 1901. CALIFORNIA. **Fresno Co.:** Pine Ridge, 15–25 Jun 1900, *H. M. Hall & Chandler 138* (HT: NY!; IT: NY!; US!).

D. laxiflora NAF **22:** 374. 1908. CALIFORNIA. **Tulare Co.:** Big Tree Canyon, 29 Jul 1891, *Coville & Funston 1355* (HT: US!; CT: NY!).

D. micropetala NAF **22:** 375. 1908. UTAH: City Creek Canyon near Salt Lake City, 9 & 12 Jun 1905, *Rydberg 6153* (HT: NY!; IT: GH! dated 12th; US! dated 12th).

D. oregana NAF **22:** 374. 1908. Columbia River, *Nutall s.n.* (HT: NY!; IT: GH!, CT: DS!).

D. pseudorupestris var. *intermedia* Mem. New York Bot. Gard. **1:** 220. 1900. MONTANA: Spanish Basin, 1896, *Flodman 597* (ST: NEB); 28 Jun 1897, *Rydberg & Bessey 4365* (ST: MONT, NEB 2 sheets, NY!).

D. pumila NAF **22:** 372. 1908. OREGON: Steen's Mts., 14 Jun 1901, *Cusick 2571* (HT: NY!; IT: GH!, US!; CT: DS!, UC).

D. viscosa NAF **22:** 372. 1908 WASHINGTON. **Okanogan Co.:** Mt. Chapaca, Aug 1897, *Elmer 564* (HT: NY!; IT: UC 2 sheets, US!).

Fragaria californica var. *franciscana* Mem. Dept. Bot. Columbia Coll. **2:** 173. 1898. CALIFORNIA. **Marin Co.:** Olena, 18 Feb & Oct 1896, *Eastwood s.n.* (NY!). The type description cites 2 specimens from California, Eastwood's from Marin Co., and one from Mt. Tamalpais collected by Brandegee. Both of these were deposited at CAS and both destroyed in the earthquake and fire of 1906.

F. crinita Mem. Dept. Bot. Columbia Coll. **2:** 171. 1898. WASHINGTON: Puget Sound, *Wilkes Expedition s.n.* (LT: US!—Rydberg, NAF **22:** 358. 1908).

F. firma Mem. Dept. Bot. Columbia Coll. **2:** 184. 1898. ARIZONA: Rincon Mts., 1891, *Neally 260* (ST: NY! no data, US!); Mogollon Mts., 21 Aug 1887, *Mearns 71* (ST: NY!); 1869, *Palmer s.n.*; 1890, *Palmer 487*;

1892, *Toumey 96*; 1892, *Wooton 91*. **COLORADO:** 1892, *Crandall s.n.*; 1873, *Coulter s.n.* **NEW MEXICO:** 1847, *Fendler 206* [type of *Potentilla ovalis* Lehm.]. **UTAH.** 1894, *Jones 5389*. **WYOMING:** Horse Creek, 9 Jun 1894, *Nelson 207* (ST: NY!).

F. insularis NAF **22:** 359. 1908. **JAMAICA.** Cinchona, 9 Apr or 4 Sep 1906, *Harris 9216* (HT: NY! on 2 sheets).

F. pauciflora Mem. Dept. Bot. Columbia Coll. **2:** 183. 1898. **MONTANA:** Gallatin Valley near Bozeman, 7 Jul 1896, *Flodman 591* (HT: NY!; IT: US!).

F. platypetala Mem. Dept. Bot. Columbia Coll. **2:** 177. 1898. **CANADA. BRITISH COLUMBIA:** Sproat, 24 Jun 1890, *Macoun s.n.* (LT: NY!—Rydberg, NAF **22:** 361. 1908; ILT: CAN).

F. porlifica Baker & Rydb. *in* Rydb., Mem. Dept. Bot. Columbia Coll. **2:** 181. 1898. **COLORADO:** Cameron Pass, 13 Jul 1896, *Baker 28* (HT: NY!).

F. pumila Mem. Dept. Bot. Columbia Coll. **2:** 181. 1898. **SOUTH DAKOTA:** Black Hills, Custer, 4 Jun 1892, *Rydberg 661* (HT: NY!; IT: NEB, US!).

F. sibbaldifolia Mem. Dept. Bot. Columbia Coll. **2:** 176. 1898. **CALIFORNIA. Tulare Co.:** North fork of the Kern River, 25 Aug 1891, *Coville & Funston 1712* (HT: US!—designated by Rydberg in herb.; IT: NY!).

F. suksdorfii NAF **22:** 361. 1908. **WASHINGTON:** Falcon Valley, 2 & 28 Jun 1883, *Suksdorf 486* (HT: NY!; IT: NY!).

F. terrae-novae Mem. Dept. Bot. Columbia Coll. **2:** 182. 1898. **CANADA.** Fortean, 2 Aug 1893, *Waghorne s.n.* (HT: NY!—designated by Rydberg in herb.).

F. truncata Mem. Dept. Bot. Columbia Coll. **2:** 177. 1898. **CALIFORNIA. Nevada Co.:** Aug 1893, *Michener & Bioletti s.n.* (LT: NY!—Rydberg, NAF **22:** 360. 1908; ILT: NY!).

F. virginiana var. *australis* Mem. Dept. Bot. Columbia Coll. **2:** 180. 1898. **NORTH CAROLINA. Stanley Co.:** Yadkin River, 20–24 Apr 1896, *Small s.n.* (LT: NY!—Rydberg, NAF **22:** 361. 1908).

F. yukonensis NAF **22:** 364. 1908. **CANADA. YUKON TERRITORY:** Lake Lebarge, 19 Jun 1899, *Tarleton 38* (HT: NY!; IT: US!).

Geum camporum NAF **22:** 403. 1913. **KANSAS:** Manhattan, 11 Aug 1892, *Norton s.n.* (HT: NY!).

G. decurrens NAF **22:** 404. 1913. **ARIZONA:** Mogollon Mts., Baker's Butte, 28 Jul 1887, *Mearns 59* (HT: NY!; IT: NY!).

G. mexicanum NAF **22:** 406. 1913. **MEXICO. [VERACRUZ]:** Mt. Orizaba, 8 Aug 1891, *Seaton 251* (HT: US!; CT: NY!).

G. perincisum NAF **22:** 405. 1913. **CANADA. ALBERTA:** Banff, Cave Ave. 21 Jul 1899, *McCalla 2074* (HT: NY!).

Holodiscus microphyllus BTBC **31:** 559. 1904. **UTAH:** Wasatch Mts., Alta, 1 Aug 1879, *Jones 1142* (HT: NY!; IT: NY!, UTC!).

Horkelia argyrocoma Mem. Dept. Bot. Columbia Coll. **2:** 144. 1898. **CALIFORNIA:** San Bernardino Mts., Bear Valley, Jun 1894, *Parish 151* (ST: US!); 1892 *Parish 2364*; 1894, *Parish 3137*; 1895, *Parish 3764*; 1876, *Parry & Lemmon 104*; 1882, *Parish & Parish 151*.

H. brownii NAF **22:** 276. 1908. **CALIFORNIA. Siskiyou Co.:** Mt. Shasta, 15–31 Jul 1897, *Brown 530* (HT: NY!; IT: UC).

H. caruifolia Rydb. *in* Howell, Fl. N.W. Amer. **1:** 181. 1898. **OREGON:** Rogue River Valley near Ashland, 17 Jul 1887, *Howell s.n.* (HT: NY!; IT: NY! 2 sheets, US!).

H. chaetophora BTBC **26:** 543. 1899. **CALIFORNIA. Tulare Co.:** Farewell Gap and Little Kern River, Aug 1896, *Purpus 1409* (LT: UC!—Rydberg, NAF **22:** 290. 1908; CT: NY!).

H. chandleri BTBC **28:** 177. 1901. **CALIFORNIA:** Mt. Goddard, 24–26 Jul 1900, *Hall & Chandler 700* (HT: NY!; IT: DS! 2 sheets, UC!, US! 2 sheets).

H. congdonis BTBC **26:** 543. 1899. **CALIFORNIA:** Casa Diablo, 22 Aug 1895, *Congdon s.n.* (HT: UC!; IT: DS!; CT: NY!).

H. flavescens Mem. Dept. Bot. Columbia Coll. **2:** 138. 1898. **CALIFORNIA:** 1875, *Lemmon 90* (HT: NY!; IT: NY! 3 sheets, US!).

H. gordonii var. *megalopetala* Mem. Dept. Bot. Columbia Coll. **2:** 152. 1898. **CALIFORNIA:** Mt. Dana, *Bolander s.n.* (HT: CAS!).

H. hispidula NAF **22:** 278. 1908. **CALIFORNIA:** White Mts., 19 Aug 1888, *Shockley 596* (HT: GH!).

H. integrifolia NAF **22:** 279. 1908. **OREGON:** Mt. Hood, 1866, *Wood 815* (HT: NY!; IT: NY! not numbered).

H. platycalyx Mem. Dept. Bot. Columbia Coll. **2:** 131. 1898. **CALIFORNIA:** Santa Barbara, 20 Feb 1893, *Wheelock s.n.* (HT: NY!; IT: NY! 2 sheets).

H. pseudocapitata Rydb. *in* Howell, Fl. N.W. Amer. **1:** 180. 1898. **CALIFORNIA:** Janesville [Tanesville in publication], 1892, *Brandegee s.n.* (not found).

H. pulchra NAF **22:** 280. 1908. **CALIFORNIA:** Mt. Shasta, 1887, *Parry s.n.* (HT: NY!).

H. scandularis Mem. Dept. Bot. Columbia Coll. **2:** 150. 1898. **CALIFORNIA:** White Mts., 18 Aug 1888, *Shockley 572* (HT: GH!; IT: CAS!, UC).

H. sericoleuca Mem. Dept. Bot. Columbia Coll. **2:** 144. 1898. **CALIFORNIA:** 1874, *Lemmon 11, 12, 71* (ST: NY! 3 sheets); 1875, *Lemmon 95* (ST: NY! 3 sheets); 1887, *Brandegee s.n.*; 1886, *Sonne 6*; **Placer Co.:** Martis Creek Valley, Jun 1888, *Sonne 353* (ST: NY! 2 sheets one not numbered).

H. tenuisecta NAF **22:** 278. 1908. **WASHINGTON:** Falcon Valley, 28 Jul 1882, *Suksdorf 2492* (HT: NY!; IT: DS!).

H. truncata NAF **22:** 274. 1908. CALIFORNIA: Ramona, collected from the Brandegee garden, 22 Aug 1895, *Chandler 5321* (HT: UC!; CT: NY).

Horkeliella NAF **22:** 282. 1908. T: *Horkelia purpurascens* S. Wats.

Ivesia alpicola Rydb. *in* Howell, Fl. N.W. Amer. **1:** 182. 1898. **WASHINGTON:** Mt. Adams, 9 Aug 1882, *Suksdorf s.n.* (ST: DS!; GH); 10 Aug 1882, *Howell s.n.* (ST: US!).

I. tweedyi NAF **22:** 288. 1908. **WASHINGTON:** Yakima Region, Aug 1883, *Tweedy s.n.* (HT: NY!).

Lachemilla bourgeaui NAF **22:** 384. 1908. **NEW MEXICO:** Santa Fe, 1865–1866, *Bourgeau 308* (HT: US!; IT: NY!).

L. orizabensis NAF **22:** 383. 1908. **MEXICO. [VERACRUZ]:** Mt. Orizaba, 1855, *Müller 1352* (HT: NY!).

L. pringlei NAF **22:** 383. 1908. **MEXICO.** **Oaxaca:** Sierra de San Felipe, 10 Sep 1894, *Pringle 4876* (HT: NY!; IT: US! 2 sheets).

L. schiedeana NAF **22:** 382. 1908. **MEXICO.** Near Angangueo, *Schiede 588* (HT: NY).

Lindleyella schiedeana NAF **22:** 259. 1908. **MEXICO.** Between Zimapán and San José del Oro, 1840, *Schiede s.n.* (HT: NY!).

Opulaster alabamensis NAF **22:** 243. 1908. **Alabama. Lee Co.:** Auburn, 5 Jun 1897, *Earle & Baker s.n.* (HT: NY!; IT: US!).

O. australis NAF **22:** 242. 1908. **North Carolina. Caldwell Co.:** Stone Mt., 31 Jul 1891, *Small & Heller 256* (HT: NY!; IT: US!).

O. bracteatus BTBC **31:** 558. 1904. **Colorado:** New Windsor, Jul 1898, *Osterhout 23* (HT: NY!).

O. cordatus NAF **22:** 242. 1908. **Idaho:** Farmington Landing, Lake Coeur d'Alene, 7 Jul 1892, *Sandberg, MacDougal & Heller 575* (HT: NY!; IT: US!).

O. glabratus BTBC **31:** 559. 1904. **Colorado:** West Spanish Peaks, 9 Jul 1900, *Rydberg & Vreeland 6027* (HT: NY!).

O. hapemanii NAF **22:** 244. 1908. **Wyoming:** Big Horn Mts., Aug 1892, *Hapeman s.n.* (HT: NY!).

O. intermedius Rydb. *in* Britton, Man. fl. n. states 492. 1901. **South Dakota:** Black Hills, French Creek, 10 miles below Custer, 22 Jun 1892, *Rydberg 653* (LT: NY!—Rydberg, NAF **22:** 242. 1908).

O. pubescens Mem. New York Bot. Gard. **1:** 205. 1900. **Montana:** Hound Creek, 31 Jul 1883, *Scribner 35* (ST: NY!); **UTAH.** Ogden, 8 May 1871, *Porter s.n.* (ST: NY!).

O. stellatus Rydb. ex Small, Fl. S.E. U.S. 513. 1903. **Georgia & Florida:** There are no specimens at NY from Georgia or Florida with this name written on the sheet in Rydberg's hand.

Oreobatus BTBC **30:** 274. 1903. T: *Rubus deliciosus* James.

Petrophytum acuminatum NAF **22:** 253. 1908. **California. Tulare Co.:** Big Arroyo, 5 Sep 1904, *Culbertson* (C. F. Baker's distribution #) *4540* (HT: NY!).

Potentilla alaskana NAF **22:** 318. 1908. **Alaska:** Kodiak Island, 11 Aug 1900, *Walpole 1201* (HT: US!).

P. amadorensis NAF **22:** 312. 1908. **California. Amador Co.:** Bear River, 30 Jul 1896, *Hansen 1946* (HT: NY!; IT: UC, US! dated 29 Jul).

P. angustata NAF **22:** 311. 1908. **California:** Upper Sacramento River, 1894, *Foster s.n.* (HT: NY!).

P. arachnoidea Dougl. ex Rydb. NAF **22:** 350. 1908.

P. argyrea NAF **22:** 341. 1908. **CANADA. Saskatchewan:** Assiniboia, Moose Jaw, 20 Jun 1896, *Macoun* (herbarim # *14441*) *s.n.* (HT: NY!).

P. atrorubens BTBC **24:** 11. 1897. **Arizona:** Mt., Graham, 1874, *Rothrock s.n.* (HT: NY!—designated by Rydberg in herb., ILT: NY! numbered *399*). In the original description only *Rothrock 399* was mentioned but the unnumbered sheet at NY has TYPE written on it in Rydberg's hand. They are probably of the same gathering.

P. atrovirens Mem. Dept. Bot. Columbia Coll. **2:** 95. 1898. **Wyoming:** Douglas Creek, 13 Aug 1896, *Nelson 2017A* (HT: NY!).

P. bakeri BTBC **31:** 560. 1904. **Colorado:** Grizzly Creek, 19 Jul 1896, *Baker s.n.* (HT: NY!; IT: UC!).

P. bicrenata BTBC **23:** 431. 1896. **New Mexico:** Agua Fria, 2 Jun 1883, *Walcott 66* (HT: US!).

P. bipinnatifida var. *platyloba* Mem. Dept. Bot. Columbia Col. **2:** 100. 1898. **Nebraska:** Williamson, 1892, *Smith & Pound 85* (HT: NEB—fide Bolick in litt. 1987).

P. bruceae NAF **22:** 342. 1908. **Oregon:** Warner Mts., Jul 1898, *Bruce 2301* (HT: NY!; CT: UC!).

P. brunnescens BTBC **28:** 173. 1901. **Wyoming:** Spread Creek, Teton Forest Reserve, Aug 1887, *Tweedy 212* (HT: NY!).

P. camporum NAF **22:** 319. 1908. **South Dakota:** Black Hills, *Pratt 93* (HT: NY!).

P. candida BTBC **24:** 6. 1897. **Nevada:** Diamond Valley, Jul 1868, *Watson 337* (HT: NY!; IT: GH!, US!, YU!).

P. cascadensis Mem. Dept. Bot. Columbia Coll. **2:** 109. 1898. **Washington: Skamania Co.:** Chiquash Mts., 14 Sep 1896, *Suksdorf 2165* (HT: NY!; IT: GH!, NY!, US! all dated 18 Aug 1892).

P. coloradensis Mem. Dept. Bot. Columbia Coll. **2:** 115. 1898. **Colorado:** South Park, 24 Jul 1872, *Porter s.n.* (LT: NY—Rydberg, NAF **22:** 339. 1908; ILT: GH!, NY!).

P. comosa NAF **22:** 316. 1908. **California:** San Bernardino Mts., Bear Valley, 19 Jun 1894, *Parish 3152* (HT: US!; CT: NY!).

P. concinna var. *divisa* BTBC **23:** 431. 1896. **CANADA.** Assiniboia, 12 Mile Lake near Wood Mt., 6 Jun 1895, *Macoun* [herbarium #*10468*] *s.n.* (LT: CAN—Rydberg, NAF **22:** 330. 1908; fide Shchepanak in litt. 1985).

P. concinnaeformis Mem. Dept. Bot. Columbia Coll. **2:** 54. 1898. **Arizona:** Mt. Agassiz, Aug 1884, *Lemmon 3294* (ST: GH!); Sep 1884, *Lemmon & Lemmon s.n.*

P. convallaria BTBC **24:** 249. 1897. **Montana:** Gallatin Valley near Bozeman, 7 Jul 1896, *Flodman 604* (HT: NY!).

P. dascia NAF **22:** 313. 1908. **Oregon:** The Dalles, 6 Jun 1869, *Harford & Dunn 1144* (HT: NY!; IT: GH no date).

P. dichroa NAF **22:** 319. 1908. **Montana:** Old Sentinel Mountain near Missoula, 12 Jun 1901, *MacDougal 185* (HT: NY!).

P. durangensis NAF **22:** 322. 1908. **MEXICO. Durango:** Quebrada Honda, 20–21 May 1906, *Palmer 223* (HT: NY!).

P. elmeri NAF **22:** 315. 1908. **California. Ventura Co.:** Jul 1902, *Elmer 4009* (HT: NY!; IT: UC!, US!).

P. etomentosa BTBC **24:** 8. 1897. **California:** 1846, *Fremont 162* (HT: NY!).

P. filipes BTBC **28:** 174. 1901. **Colorado:** Wahatoya Canyon, 13 Jun 1900, *Rydberg & Vreeland 6039* (HT: NY!; IT: NEB).

P. flabelliformis var. *ctenophora* BTBC **24:** 7. 1897. "Both forms [the other being typical *flabelliformis*] are fairly common from Wyoming and California to British Columbia and Saskatchewan." (Not found).

P. flavovirens NAF **22:** 306. 1908. **MEXICO. Lower**

CALIFORNIA: 8 Jul 1885, *Orcutt s.n.* (HT: NY!; IT: UC!).

P. glabrella Mem. Dept. Bot. Columbia Coll. **2:** 94. 1898. Rocky Mts., Silver City, 4 Aug 1885, *Macoun* (herbarium #'s *628, 8004*) *s.n.* (ST: both CAN); **CANADA.** SASKATCHEWAN: Prince Albert, 8 Jul 1896, *Macoun* (herbarium *#12567*) *s.n.* (ST: CAN); Cypress Hills, 1 Aug 1880, *Macoun* (herbarium #'s *628, 1445, 8010*) *s.n.* (ST: all CAN). Typified to Rocky Mts. by Rydberg, NAF **22:** 350. 1908.

P. goldmanii Painter ex Rydb., NAF **22:** 314. 1908. **MEXICO.** OAXACA: Mt. Zempoaltepec, 10 Jul 1894, *E. W. Nelson & Goldman 691* (HT: US!).

P. grosseserrata NAF **22:** 312. 1908. WASHINGTON: 1899, *Vasey 322* (HT: NY!; IT: US!).

P. hallii BTBC **28:** 176. 1901. CALIFORNIA. **Fresno Co.:** Pine Ridge, 15–25 Jun 1900, *H. M. Hall & Chandler 182* (HT: NY!; IT: NY!, UC, US!).

P. hassei NAF **22:** 329. 1908. CALIFORNIA. **San Diego Co.:** Strawberry Valley, 12 Jul 1892, *Hasse 5696* (HT: NY!).

P. horrida BTBC **28:** 173. 1901. **MEXICO.** CHIHUAHUA: Sierra Madres near Colonia García, 27 Jun 1899, *Townsend & Barber 16* (HT: NY!; IT: GH!, NY!, UC 2 sheets, US! 2 sheets).

P. intermittens NAF **22:** 318. 1908. COLORADO: Cameron Pass, 13 Jul 1896, *Baker 25* (HT: NY!).

P. kelseyi NAF **22:** 306. 1908. MONTANA: Helena, 9 Jun 1891, *Kelsey s.n.* (HT: NY!).

P. klamathensis NAF **22:** 343. 1908. OREGON: Near Fort Klamath, 7 Aug 1894, *Leiberg 660* (HT: US!; IT: GH!, NY!, UC; CT: NY!).

P. langeana BTBC **28:** 179. 1901. **GREENLAND.** Amaralik Fjord, Jul 1831, *Vahl s.n.* (HT: NY!).

P. lasia NAF **22:** 314. 1908. CALIFORNIA: San Bernardino Mts., Bear Valley, 19 Jun 1894, *Parish 3252* (HT: NY!; IT: NY!).

P. lasiodonta NAF **22:** 351. 1908. **CANADA.** ALBERTA: Near Calgary, 21 Jul 1897, *Macoun* (herbarium *#16716*) *s.n.* (HT: NY!; IT: CAN).

P. lateriflora BTBC **23:** 261. 1896. ARIZONA: 1877, *Palmer 143* (ST: NY!); rim of Grand Canyon, 12 Jul 1892, *Wooton s.n.* (ST: NY!). CALIFORNIA: Wood Canyon, Grapevine Mts., 1891, *Coville & Funston 1763* (ST: NY!); *Wilkes expedition 1647*; 1875, *Lemmon 85*; 1894, *Parish 3153*; 1881, *Jones 2700.* IDAHO: Scoopers Spring, 21 Jul 1892, *Mulford s.n.* (ST: NY!). MONTANA: 1895, *Moore s.n.*; Deer Lodge, 8 Jul 1895, *Rydberg s.n.* (ST: NY!). NEVADA: Unionville, 1867, *Watson 324* in part. OREGON: 1875, *Bartholf 29*; 1881, *Howell s.n.* UTAH: 1859, *Engelmann s.n.* **CANADA.** BRITISH COLUMBIA: 30 Jul 1889, *Macoun s.n.* (ST: NY!); Assiniboia, Cardston, Lees Creek, 25 Jul 1895, *Macoun* [herbarium *#10482*] *s.n.* (ST: NY!).

P. leptophylla NAF **22:** 310. 1908. **MEXICO.** COAHUILA: Caracol Mts., Feb–Oct 1880, *Palmer 327* (HT: US!; IT: GH!, NY!).

P. leucocarpa NAF **22:** 307. 1908. WASHINGTON: Near Bingen, 8 May 1891, *Suksdorf 2011* (HT: NY!).

P. litoralis BTBC **23:** 264. 1896. **CANADA.** LABRADOR: 1882, *Allen s.n.* NEWFOUNDLAND: 1895, *Waghorne 8.* QUEBEC: Ste. Anne des Monts, Gaspé, 10 Jul 1881, *Allen s.n.* (ST: NY!). MAINE: Cape Elizabeth, 18 Jul 1861, *Boott s.n.* (ST: NY!); *Fernald s.n.*; NEW HAMPSHIRE: Isle of Shoals, *Oakes & Robbins s.n.* (ST: NY!, US!).

P. longiloba NAF **22:** 317. 1908. MONTANA: Lo-Lo, 29 May 1897, *Elrod 110* (HT: NY!).

P. longipedunculata Mem. Dept. Bot. Columbia Coll. **2:** 39. 1898. OREGON: Monmouth, 28 May 1893, *S. Howell s.n.* (HT: Herb. Experimental Station at Pullman, Washington, not found in WS by Gilmartin in litt. 1984; IT: US!).

P. lupina NAF **22:** 340. 1908. WYOMING: Big Horn Mts., headwaters of Clear Creek and Crazy Woman River, 20 Jul–15 Aug 1900, *Tweedy 3215* (HT: NY!).

P. luteosericea Mem. Dept. Bot. Columbia Coll. **2:** 101. 1898. **MEXICO.** LOWER CALIFORNIA: 1893, *Brandegee s.n.* (HT: CAS!; IT: UC dated 19 May).

P. macounii Mem. Dept. Bot. Columbia Coll. **2:** 101. 1898. **CANADA.** ALBERTA: Crow's Nest Pass, 4 Aug 1897, *Macoun* (herbarium *#16709*) *s.n.* (LT: NY!—Rydberg, NAF **22:** 338. 1908; ILT: GH! dated Aug 1897).

P. macropetala NAF **22:** 313. 1908. OREGON. **Tillamook Co.:** Tillamook, 27 Jun 1894, *Lloyd s.n.* (HT: NY!).

P. michoacana Mem. Dept. Bot. Columbia Coll. **2:** 44. 1898. **MEXICO.** MICHOACÁN: Morelia, 21 Jul 1892, *Pringle 5291* (HT: GH!).

P. millefolia BTBC **23:** 433. 1896. CALIFORNIA: 1875, *Lemmon 86* (HT: NY! on five sheets; ?IT: UC! 2 sheets, US!).

P. minutifolia BTBC **23:** 339. 1896. COLORADO: Pikes Peak, 24 Aug 1895, *Canby s.n.* (HT: US!; IT: GH!, UC both dated 27 Aug).

P. modesta NAF **22:** 331. 1908. UTAH: Mt. Barette, 26 Jul 1905, *Rydberg & Garrett 7261* (HT: NY!; IT: UC!).

P. nelsoniana NAF **22:** 344. 1908. WYOMING: La Plata Mines, 23 Aug 1895, *Nelson 1819* (HT: NY!; IT: GH!, UTC! both dated 21 Aug).

P. newberryi var. *arenicola* Mem. Dept. Bot. Columbia Coll. **2:** 112. 1898. WASHINGTON: Wallula, 1896, *Howell s.n.* OREGON. **Lake Co.:** North from Button Springs, 3 Jul 1894, *Leiberg 382* (ST: UC 2 sheets, US! 2 sheets).

P. nipharga NAF **22:** 332. 1908. **CANADA.** [NORTHWEST TERRITORIES]: Fort Good Hope on the Mackenzie River, 1861–1862, *Onion s.n.* (HT: NY! on 2 sheets).

P. oaxacana Mem. Dept. Bot. Columbia Coll. **2:** 206. 1898. **MEXICO.** OAXACA: Cuatro Venados, 27 Jun 1894, *L. C. Smith 47* (HT: GH!).

P. oblanceolata Mem. Dept. Bot. Columbia Coll. **2:** 53. 1898. **MEXICO.** CHIHUAHUA: Aug–Nov 1885, *Palmer* (HT: US!).

P. obovatifolia BTBC **28:** 175. 1901. **MEXICO.** *Pringle 6890* (HT: NY!; IT: GH!, NY!, US! all labeled Sierra de Pachuca, 4 Aug 1898).

P. parishii NAF **22:** 313. 1908. CALIFORNIA. **San Diego Co.:** Descanso, 7 Jun 1897, *Parish 4523* (HT:NY!).

P. paucijuga NAF **22:** 348. 1908. UTAH: La Sal Mts.,

1899, *Purpus 251* in part (HT: US not found; IT: UC; CT: NY!).

P. pecten NAF **22:** 315. 1908. **Montana:** Bridger Mts., 10 Jun 1897, *Rydberg & Bessey 4377* (HT: NY!).

P. pectinisecta BTBC **24:** 7. 1897. **Utah:** Salt Lake City, 10 Jun 1880, *Jones 1765*(LT: NY!—Rydberg, NAF **22:** 317. 1908; ILT: BRY, RM).

P. perdissecta NAF **22:** 327. 1908. **Montana:** Madison Range, Spanish Peaks, 14 Jul 1896, *Flodman 572* (HT: NY!; IT: US!).

P. permollis BTBC **28:** 175. 1901. **Washington. Whitman Co.:** Endicot, 8 Jun 1898, *Elmer 1030* [*1830* in publication] (HT: NY!; IT: US dated Jun 1898).

P. proxima NAF **22:** 339. 1908. **Utah:** Divide between Sevier and Beaver Rivers, near Belknap, 28 Jul 1905, *Rydberg & Carlton 7369* (HT: NY!).

P. pseudorupestris BTBC **24:** 250. 1897. **Montana:** Little Belt Mts., Long Baldy, 19 Aug 1896, *Flodman 598* (HT: NY!; IT: US!).

P. pseudosericea Mem. Dept. Bot. Columbia Coll. **2:** 98. 1898. Rocky Mts., *Nuttall s.n.* (HT: NY!; IT: GH!).

P. ramulosa BTBC **23:** 430. 1896. **Arizona:** 1881, *Lemmon 399*; Flagstaff, 7 Jun 1883, *Rusby s.n.* (ST: NY!).

P. rectiformis NAF **22:** 312. 1908. **Washington:** Pullman, 8 Jul 1896, *Elmer 69* (HT: NY!).

P. rhomboidea BTBC **23:** 248. 1896. **Oregon:** Deer Creek Mts., 5 Jul 1887, *Howell s.n.* (LT: NY!— Rydberg, NAF **22:** 376. 1908).

P. rosulata BTBC **26:** 542. 1899. **California:** Colorado Desert, 29 Palms, 7 Jun 1898, *Alverson s.n.* (HT: UC!).

P. rubripes BTBC **33:** 143. 1906. **Colorado:** Pikes Peak, 25 Jun 1896, *Biltmore* (Colorado Expedition #) *1425* (HT: NY!).

P. sanguinea NAF **22:** 324. 1908. **Arizona:** Walnut Canyon, 23 Jul 1898, *MacDougal 331* (HT: NY!; IT: GH!, UC, US!).

P. saximontana BTBC **23:** 399. 1896. **Colorado:** 1873, *Wolf 366* (HT: GH!; IT: US! with *Wolf & Rothrock 366* as the collectors).

P. sierrae-blancae Wooton & Rydb. *in* Rydb., Mem. Dept. Bot. Columbia Coll. **2:** 57. 1898. **New Mexico:** White Mts., 6 Aug 1897, *Wooton 469* (HT: NY!; IT: NY!, US!).

P. staminea Mem. Dept. Bot. Columbia Coll. **2:** 67. 1898. **MEXICO. Chiapas** etc., 1864–1870, *Ghiesbrecht 131 & 681* (ST: GH!, NY! both numbered *681*).

P. subcoricea Mem. Dept. Bot. Columbia Coll. **2:** 63. 1898. **MEXICO. México:** Sierra de las Cruces, 11 Sep 1892, *Pringle 5258* (HT: GH!).

P. subjuga BTBC **23:** 397. 1896. **Colorado:** Near Empire, 28 Jul & 6 Sep 1892, *Patterson 192* (HT: NY!; IT: GH!, MO!, UC).

P. subquinata var. *pedersenii* BTBC **28:** 182. 1901. **GREENLAND.** Disco, Vaigat Assuk, 25 Jul 1898, *Pedersen 470* (HT: NY!; IT: CAN).

P. subvillosa NAF **22:** 316. 1908. **California. Alpine Co.:** Carson Spur, 1892, *Hansen 297* (HT: NY!).

P. tenerrima BTBC **23:** 398. 1896. **Colorado:** Bergen's Park, 1874, *Brandegee s.n.* (HT: NY!; IT: GH! labeled Fremont Co., MO! numbered *950,* UC).

P. townsendii BTBC **28:** 174. 1901. **MEXICO. Chihuahua:** Sierra Madre, near Colonia García, 18 Jul 1899, *Townsend & Barber 105* (HT: NY!; IT: GH!, NY! 2 sheets, UC, US! 2 sheets).

P. versicolor NAF **22:** 344. 1908. **Oregon:** Grayhart Buttes, 9 Aug 1896, *Coville & Leiberg 307* (HT: US!; CT: NY!).

P. viridescens Mem. Dept. Bot. Columbia Coll. **2:** 69. 1898. **CANADA. Alberta:** Braggs Creek, 9 Jul 1897, *Macoun* (herbarium *#16746*) *s.n.* (HT: NY!—designated by Rydberg in herb.; IT: CAN).

P. viridior NAF **22:** 342. 1908. **Colorado:** Estes Park, 12 Jul 1897, *Osterhout 12* (HT: NY!).

P. vreelandii NAF **22:** 325 1908. **Montana:** Sperry Glacier, 6 Aug 1901, *Vreeland 1092* (HT: NY!; IT: CAN!).

P. wheeleri var. *viscidula* BTBC **23:** 429. 1896. **Arizona:** Santa Rita Mts., 3 May 1881, *Pringle s.n.* (HT: NY! on 3 sheets; IT: US!).

Prunus corymbulosa Mem. New York Bot. Gard. **1:** 226. 1900. **Montana:** Bridger Mts., 18 Jun 1897, *Rydberg & Bessey 4437* (HT: NY! on 2 sheets).

P. rugosa Brittonia **1:** 90. 1931. **Kansas:** Sandhills south of Cimarron River, north of Elkhart, 12 Jul 1929, *Rydberg & Imler 906* (HT: NY! on 2 sheets).

Rosa arizonica NAF **22:** 516. 1918. **Arizona:** Flagstaff, 15 Jun 1898, *MacDougal 110* (HT: NY!; IT! NY!, US!).

R. aucuparia NAF **22:** 501. 1918. **Missouri:** Dumal, 6 Jul 1919, *Bush 5866* (HT: NY!).

R. bakeri Fl. Colorado 190, 191. 1906. **Colorado. Routt Co.:** Four Mile Hill, 2 Jul 1896, *Baker s.n.* (HT: NY!; IT: NY!).

R. bicknellii NAF **22:** 500. 1918. **Massachusetts:** Nantucket Island, Coskaty, 14 Aug 1906, *Bicknell s.n.* (HT: NY!).

R. bidenticulata NAF **22:** 518. 1918. **California. Shasta Co.:** Castella, 25 [24 in publication] Jul 1912, *Eastwood 1386* [*1389* in publication] (HT: GH!).

R. brachycarpa BTBC **44:** 71. 1917. **California:** Temescal Canyon, near Elsinore, 23 May 1892, *McClatchie s.n.* (HT: NY!).

R. brownii BTBC **44:** 70. 1917. **California. Siskiyou Co.:** North side of Mt. Shasta, 11–16 Jun 1897, *Brown 349* (HT: NY!; IT: NY! dated 1–15 Jun, US!).

R. bushii NAF **22:** 506. 1918 **Missouri:** Courtney, 2 Jul 1909, *Bush 5860* (HT: NY!).

R. butleri NAF **22:** *506. 1918.* **Montana:** Helena, 28 Jun 1908, *Butler 796* (HT: NY!).

R. chrysocarpa BTBC **44:** 74. 1917. **Utah:** Allen Canyon, southwest of Abajo Mt., 30–31 Jul 1911, *Rydberg & Garrett 9302* (HT: NY!).

R. collaris Fl. Rocky Mts. 441, 1062. 1917 [1918]. **Idaho:** Pinehurst, 17 Aug 1911, *Macbride 1676* (HT: NY!; IT: NY!).

R. columbiana NAF **22:** 514. 1918. **Idaho. Latah Co.:** Valley of the Little Potlatch River, 13 Jun 1892, *Sandberg, MacDougal & Heller 381* (HT: HY!).

R. conjuncta NAF **22:** 505. 1918. **Missouri. Atchison Co.:** 23 Aug 1893, *Bush 101* (HT: NY!).

R. copelandii E. L. Greene ex Rydb., BTBC **44:** 80. 1917. CALIFORNIA. **Siskiyou Co.:** Mt. Eddy, 8 Sep 1903, *Copeland* (C. F. Baker's #) *3875* (HT: NY!; IT: US!).

R. corymbiflora NAF **22:** 519. 1918. CALIFORNIA. **Shasta Co.:** Pitt to Baird, 25 Jul 1912, *Eastwood 1404* (HT: GH!; IT: NY!).

R. davyi BTBC **44:** 76. 1917. CALIFORNIA. **Santa Clara Co.:** Saratoga, Sep 1893, *Davy 263* HT: NY!).

R. dudleyi BTBC **44:** 73. 1917. CALIFORNIA. **Fresno Co.:** Converse Basin, near Booles Home, 16 Oct 1900, *Dudley 3388* (HT: DS!; IT: CAS!).

R. eastwoodiae NAF **22:** 527. 1918. CALIFORNIA. **Siskiyou Co.:** Sisson, 4 Sep 1912, *Eastwood 2100* (HT: GH!; IT: NY!).

R. floridana Rydb. ex Small, Shrubs of Florida 27, 133. 1913. FLORIDA: Near Jacksonville, 10 May & 2 Nov 1894, *Curtiss 4754* [*4755* in publication] (HT: NY!).

R. granulifera NAF **22:** 517. 1918. ARIZONA: Holbrook, 10 Jul 1896, *Zuck s.n.* (HT: NY! on 2 sheets; IT: US!).

R. greenei BTBC **44:** 71. 1917. CALIFORNIA: Santa Cruz Island, Jul & Aug 1886, *Greene s.n.* (HT: US!; CT: NY!).

R. johnstonii NAF **22:** 521. 1918. CALIFORNIA. **San Bernardino Co.:** Upland, 4 Jul 1918, *Johnston 2050* (ST: NY!); 25 Aug 1918, *Johnston 2130* (ST: NY!).

R. nanella NAF **22:** 497. 1918. MASSACHUSETTS: Chappaquiddick Island, 29 Sep 1915, *Bicknell s.n.* (HT: NY!; IT: NY! numbered *4773*).

R. obtusiuscula NAF **22:** 497. 1918. TENNESSEE. **Cocke Co.:** French Broad River, between Paint Rock and Del Rio, 25 Aug 1897, *Kearney 695* (HT: NY!).

R. oligocarpa NAF **22:** 532. 1918. CALIFORNIA. **Shasta Co.:** Goose Valley, 29 Jun & 11 Jul 1912, *Eastwood 945* (HT: GH!; IT: NY!).

R. oreophila BTBC **31:** 561. 1904. COLORADO. **Routt Co.:** Four Mile Hill, 22 Jul 1896, *Baker s.n.* (HT: NY!; IT: NY!).

R. palmeri NAF **22:** 502. 1918. MISSOURI: Carthage, 13 Aug 1911, *Palmer 3438* [*3428* in publication] (HT: GH!; IT: GH!).

R. palustriformis NAF **22:** 516. 1918. WISCONSIN: Green Bay, 1 Jun 1895, *Schuette s.n.* (HT: GH!; IT: NY! 2 sheets).

R. pandorana E. L. Greene ex Rydb., Fl. Colorado 190, 191. 1906. COLORADO: Pandora, 10 Aug 1901, *Baker 751* (HT: NY!).

R. petiolata NAF **22:** 501. 1918. MISSOURI. **Clark Co.:** 27 Aug 1892, *Bush s.n.* (HT: MO—fide Wolf in litt. 1984).

R. pilifera BTBC **44:** 80. 1917. CALIFORNIA: San Francisco, *Bolander s.n.* (HT: NY!).

R. pringlei BTBC **44:** 79. 1917. CALIFORNIA. **Siskiyou Co.:** 6 Sep 1882, *Pringle s.n.* (HT: NY!; IT: US!).

R. puberulenta Fl. Rocky Mts. 443, 1062. 1917 [1918]. UTAH: Montezuma Canyon east of Monticello, 13 Aug 1911, *Rydberg & Garrett 9705* (HT: NY!).

R. pyrifera Fl. Rocky Mts. 445, 1062. 1917 [1918]. IDAHO: Lake Pend d'Oreille, 7 Aug 1892, *Sandberg, MacDougal & Heller 871* (HT: NY!; IT: US!).

R. rotundata BTBC **44:** 76. 1917. NEVADA: Mountains west of Franktown, 2 Jul 1912, *Heller 10520* (HT: NY!; IT: DS!, RENO!, US!).

R. salictorum BTBC **44:** 77. 1917. NEVADA: Gold Creek, 25 Jul 1912, *Nelson & Macbride 2113* (HT: NY!; IT: NY!).

R. santae-crucis BTBC **44:** 73. 1917. CALIFORNIA: Island of Santa Cruz, Jul & Aug 1886, *Greene s.n.* (HT: NDG—fide Hellenthal in litt. 1984).

R. standleyi NAF **22:** 527. 1918. NEW MEXICO. **San Miguel Co.:** Pecos, 18 Aug 1908, *Standley 5043* (HT: NY!).

R. subglauca NAF 22: 503. 1918. **CANADA.** SASKATCHEWAN: 6 miles east of Battle River, 15 Aug 1906, *Macoun & Herriot* (herbarium *#70943*) *s.n.* (HT: NY!; IT: CAN).

R. subserrulata NAF **22:** 500. 1918. MISSOURI: Swan, 2 Jun 1899, *Bush 42* (HT: NY!).

R. texarkana NAF **22:** 503. 1918. ARKANSAS. **Miller Co.:** Texarkana, 24 Jul 1896, *Eggert s.n.* (HT: MO—fide Wolf in litt. 1984; IT: MO; CT: NY!).

R. treleasei NAF **22:** 503. 1918. LOUISIANA: Alden Bridge, 30 Jun 1898, *Trelease s.n.* (HT: MO on 2 sheets—fide Wolf in litt. 1984; CT: NY!).

R. underwoodii BTBC **31:** 560. 1904. COLORADO: Hills about Box Canyon west of Ouray, 8 Sep 1901, *Underwood & Selby 122* (HT: NY!).

Rubacer BTBC **30:** 274. 1903. T: *Rubus odoratus* L.

Rubus aboriginum NAF **22:** 473. 1913. TEXAS: Big Sandy, 7 Apr 1902 [1862 in publication], *Reverchon 2988* (HT: MO—fide Wolf in litt. 1984).

R. alnifolius NAF **22:** 457. 1913. **MEXICO.** VERACRUZ: San Miguel near Jalapa, 17–22 May 1899, *Rose & Hough 4295* (HT: US!; CT: NY!).

R. amplior NAF **22:** 456. 1913. **GUATEMALA.** BAJA VERAPAZ: Santa Rosa, Sep 1888, *von Türckheim 1424* (HT: US!; CT: NY!).

R. brainerdi NAF **22:** 467. 1913. VERMONT: Weybridge, Otter Creek, 16 Jun 1899 [1896 in publication] *Brainerd s.n.* (ST: NY!); 12 Aug 1896, *Brainerd s.n.* (ST: NY!).

R. carpinifolius Rydb. *in* Small, Fl. S.E. U.S. 519, 1331. 1903. LOUISIANA: Alexandria, 22 May 1899, *Ball 428* (HT: NY!).

R. carolinianus NAF **22:** 447. 1913. NORTH CAROLINA. **Swain Co.:** Andrews Bald, 25 Jul 1891, *Beardslee & Kofoid s.n.* (HT: NY!).

R. cymosus NAF **22:** 470. 1913. **MEXICO. [MÉXICO]:** Tlalpam, valley of Mexico, 18 Jul 1901, *Rose & Hay 5540* (HT: US!; IT: NY!).

R. eastwoodianus NAF **22:** 460. 1913. CALIFORNIA: Mt. Tamalpais, 15 Jun 1905, *Rydberg 6238* (HT: NY! on 3 sheets).

R. franciscanus NAF **22:** 441. 1913. CALIFORNIA: Punta de los Reyes, 1853–1854, *Bigelow s.n.* (LT: NY!—Hoewll et al., Wasmann J. Biol. **16:** 85. 1958).

R. helleri NAF **22:** 460. 1913. WASHINGTON. **Chehalis Co.:** Montesano, 30 Jun 1898, *Heller & Heller 3990* (HT: NY!; IT: US!).

R. laxus NAF **22:** 454. 1913. **GUATEMALA.** SANTA ROSA: Zamorora, May 1893, *Heyde & Lux 4474* (HT: US!; CT! NY!).

R. lucidus Rydb. ex Small, Shrubs of Florida 26, 133. 1913. FLORIDA: Dunedin, 17 Apr 1900, *Tracy 6855* (HT: NY!).
R. nelsonii NAF **22:** 455. 1913. **MEXICO.** OAXACA: Mt. Zempoaltepec, 5–13 Jul 1894, *E. W. Nelson 580* (HT: US!; IT: NY!).
R. oligospermus Thornber ex Rydb., NAF **22:** 470. 1913. ARIZONA: Apr 1881, *Pringle s.n.* (HT: US!; IT: NY! as Santa Catalina Mts.).
R. ostryifolius Rydb. *in* Britton, Man. fl. n. states 497. 1901. NEW JERSEY. **[Monmouth Co.]:** Atlantic Highlands, Aug 1892, *Britton s.n.* (HT: not found at NY).
R. palmeri NAF **22:** 456. 1913. **MEXICO.** DURANGO: San Ramón, 21 Apr–18 May 1906, *Palmer 78* (HT: NY!; IT: US!).
R. persistens Rydb. *in* Small, Fl. S.E. U.S. 519, 1331. 1903. MISSISSIPPI: Biloxi, 22 Mar 1898, *Tracy 4108* (HT: NY!; IT: BH).
R. pittieri NAF **22:** 454. 1913. **COSTA RICA.** Abejonal, 4 May 1890, *Pittier 2281* (HT: US!; CT: NY!).
R. rhodophyllus Rydb. *in* Small, Fl. S.E. U.S. 518, 1331. 1903. MISSISSIPPI: Point St. Martin, 21 Mar 1898, *Tracy 4107* (HT: NY!).
R. rubrisetus Rydb. *in* Britton, Man. fl. n. states 497. 1901. LOUISIANA: New Orleans, Feb 1885, *Poggenburg s.n.* (HT: NY!).
R. tuerckheimii NAF **22:** 457. 1913. **GUATEMALA.** ALTA VERAPAZ: Cobán, Dec 1902, *von Türckheim 8387* (HT: US!; CT: NY!).
R. verae-crucis NAF **22:** 450. 1913. **MEXICO.** VERACRUZ: San Miguel del Soldado, 20 Apr 1899, *Pringle 8177* (HT: NY!; IT: US!).
Sanguisorba menziesii NAF **22:** 387. 1908. ALASKA: Short Bay, 8 Aug 1895, *Howell 1620* (HT: NY!).
Sericotheca concolor NAF **22:** 264. 1908. NEVADA: Mt. Davidson, 1865, *Torrey 134* (HT: NY!; IT: NY!).
S. franciscana NAF **22:** 262. 1908. CALIFORNIA: San Leandro, Jun 1888, *Underwood s.n.* (HT: NY!).
S. obovata NAF **22:** 264. 1908. CALIFORNIA. **Placer Co.:** Truckee River, 17 Jul 1886, *Sonne s.n.* (HT: NY!).
S. pachydisca NAF **22:** 263. 1908. **MEXICO.** Tacubaya, 1865–1866, *Bourgeau 267* (HT: US!; CT: NY!).
S. schaffneri NAF **22:** 264. 1908. **MEXICO.** SAN LUIS POTOSI: 1879, *Schaffner 451* (HT: NY!; IT: NY!, US!).
S. velutina NAF **22:** 265. 1908. **MEXICO.** OAXACA: Sierra de San Felipe, 25 Sep 1894, *C. L. Smith 821* (HT: NY!; IT: US!).
Sibbaldiopsis Mem. Dept. Bot. Columbia Coll. **2:** 187. 1898. T: *Potentilla tridentata* Soland.
Sieversia macrantha Kearney ex Rydb., NAF **22:** 412. 1913. ALASKA: Kukah Bay, 1–5 Jul 1899, *Coville & Kearney 1696* (HT: US!).
Sorbus angustifolia Fl. Rocky Mts. 448, 1062. 1917 [1918]. IDAHO: Cedar Mt., Jul 1898, *Piper s.n.* (HT: NY!).
Spiraea helleri NAF **22:** 248. 1908. CALIFORNIA. **Nevada Co.:** Donner Pass, 27 Jul 1903, *Heller 7022* (HT: NY!; IT: US! 2 sheets).
S. roseata NAF **22:** 250. 1908. IDAHO: Petitt Lake, 13 Aug 1895, *Evermann 304* (HT: US!; CT: NY!).
S. subcanescens NAF **22:** 252. 1908. SOUTH CAROLINA: Anderson, 1885, *Gibbes s.n.* (HT: NY!).
S. subvillosa NAF **22:** 251 1908. OREGON: Cascades of the Columbia, Jun 1886, *Howell s.n.* (HT: NY!).
S. tomentulosa NAF **22:** 251. 1908. WASHINGTON: Falcon Valley, 7 Jul–Sep 1881, *Suksdorf 5* (HT: US!).
Vauquelinia angustifolia NAF **22:** 260. 1908. **MEXICO.** CHIHUAHUA: Santa Eulalia Mts., 27 May 1885, *Pringle 5* (HT: NY!; IT: NY! 2 sheets, UC, US!).
V. latifolia Rydb. ex Standley, Contr. U.S. Natl. Herb. **23:** 323. 1922. **MEXICO.** TAMAULIPAS: Near Miquihuana, 10 Jun 1898, *E. W. Nelson 4481* (HT: US!; CT: NY!).
Zygalchemilla NAF **22:** 385. 1908. T: *Alchemilla pinnata* Ruíz & Pavón.

RUBIACEAE

Galiumboreale var. *linearifolium* Mem. New York Bot. Gard. **1:** 375. 1900. NEBRASKA: Canyon south of Pumpkin Seed Valley, 11 Jul 1891, *Rydberg 134* (HT: NY!; IT: NEB 2 sheets).
G. columbianum Fl. Rocky Mts. 808, 1066, 1917 [1918]. IDAHO. **Kootenai Co.:** Wiessner's Peak, 8 Jul 1892, *Sandberg, Heller & MacDougal 600* (HT: NY!).
G. filipes Fl. Rocky Mts. 809, 1066. 1917 [1918]. UTAH: City Creek Canyon, 4 Jul 1883, *Leonard 166/65* (HT: NY!).
G. flaviflorum A. A. Heller ex Rydb., Fl. Colorado 321, 322. 1906. NEW MEXICO: Santa Fe Canyon, 9 miles east of Santa Fe, 7 Jul 1897, *Heller & Heller 3823* (HT: NY!; IT: BKL).

RUPPIACEAE

Ruppia pectinata Mem. New York Bot. Gard. **1:** 18. 1900. **[**WYOMING**]:** Yellowstone Park, Aug 1884, *Tweedy 229* (HT: YU!).

SALICACEAE

Populus acuminata BTBC **20:** 50. 1893. NEBRASKA. **Scotts Bluff Co.:** 25 Jul 1891, *Rydberg 372* (HT: NY!—designated by Rydberg in herb.; IT: US! labeled Carter Canyon and dated 26 Jul).
Salix atra Bull. New York Bot. Gard. **1:** 272. 1899. **CANADA.** LABRADOR: Ford's Harbor, 31 Jul 1884, *Bell 18823* (LT: CAN—Argus, Contr. Gray Herb. **196:** 64. 1965).
S. cyclophylla Bull. New York Bot. Gard. **1:** 274. 1899. ALASKA: St. Paul's Island, 5 Jul 1897, *Macoun* (herbarium *#16645*) *s.n.* (HT: NY!).
S. dodgeana Bull. New York Bot. Gard. **1:** 277. 1899. WYOMING: Yellowstone Park, Electric Peak, 18 Aug 1897, *Rydberg & Bessey s.n.* (LT: NY!—Cronquist, Univ. Wash. Publ. Biol. **17(2):** 49. 1964; ILT: MONT, RM!, US! all numbered 3921).
S. flava BTBC **28:** 273. 1901. WYOMING: Green River, 25 Jun 1895, *Rydberg s.n.* (HT: NY!).

S. labradorica Bull. New York Bot. Gard. **1:** 274. 1899. **CANADA. Labrador:** Turner's Head, 8 Jun 1892, *Waghorne 36* (HT: NY!—designated by Rydberg in herb.).

S. longistylis Rydb. *in* Britton & Rydb., Bull. New York Bot. Gard. **2:** 163. 1901. **CANADA. Yukon:** Dawson, mouth of Klondike, 30 May 1899, *Williams s.n.* (HT: NY!).

S. luteosericea Rydb. *in* Britton, Man. fl. n. states 316. 1901. **Nebraska. Banner Co.:** Lawrence Fork, 8 Jul 1891, *Rydberg 368* (HT: NY!).

S. macounii Bull. New York Bot. Gard. **1:** 269. 1899. **CANADA. Quebec:** Anticosti, Ellis Bay, 1883, *Macoun* (herbarium *#18830) s.n.* (HT: CAN—fide Shchepanek in litt. 1985).

S. niphoclada Bull. New York Bot. Gard. **1:** 272. 1899. **CANADA. Northwest Territories:** Mackenzie River, 30 miles north of Arctic Circle, 1892, *Taylor s.n.* (HT: CAN—fide Shchepanek in litt. 1984; IT: US! dated 18 Jul).

S. pachnophora BTBC **31:** 403. 1904. **Colorado:** Chambers Lake, 30 Jul 1899, *Crandall 4083* (HT: NY!; IT: CS, US!).

S. padifolia BTBC **28:** 272. 1901. **Colorado:** Turkey Creek and tributaries, 8–9 Jun 1900, *Rydberg & Vreeland 6389* (HT: NY!).

S. palaeoneura Bull. New York Bot. Gard. **1:** 267. 1899. **Alaska:** Point Barrow, 11 Jun, *Murdock b* (HT: GH!).

S. perrostrata Rydb. *in* Britton & Rydb., Bull. New York Bot. Gard. **2:** 163. 1901. **South Dakota:** Black Hills, Hermosa, 23 Jun 1892, *Rydberg 1018* (HT: NY!; IT: US! labeled Hot Springs and dated 9 Aug).

S. sandbergii BTBC **39:** 304. 1912. **Idaho:** Valley of Hatwai Creek, 28 Apr 1892, *Sandberg, MacDougal & Heller 71* (HT: NY!; IT: US!).

S. saximontana Bull. New York Bot. Gard. **1:** 261. 1899. **Colorado:** Gray's Peak, Aug 1895, *Rydberg s.n.* (HT: NY!).

S. seemannii Rydb. *in* Britton & Rydb., Bull. New York Bot. Gard. **2:** 164. 1901. **CANADA. Yukon:** Dawson, 11 Jun 1899, *Williams s.n.* (HT: NY!).

S. squamata Rydb. *in* Britton, Man. fl. n. states 317. 1901. **New Jersey. Sussex Co.:** Sparta Junction, 17 Sep 1887, *Porter & Britton s.n.* (HT: NY!).

S. stenophylla BTBC **28:** 271. 1901. **Colorado:** Cuchara River below La Veta, 28 May 1900, *Rydberg & Vreeland 6393* (HT: NY!).

S. wyomingensis BTBC **28:** 271. 1901. **Wyoming:** Big Horn Mts., headwaters of Clear Creek & Crazy Woman River, 20 Jul–15 Aug 1900, *Tweedy 3434* (HT: NY!).

SANTALACEAE

Comandra californica Eastw. ex Rydb., Fl. Rocky Mts. ed 2, 1138. 1922 [1923]. **California. Tulare Co.:** south Fork of Kaweah River, 20 Jul 1904, *Culbertson 4479* (LT: CAS!—Piehl, Mem. Torrey Bot. Club **22:** 65. 1965; ILT: CAS!, GH, MO, POM).

C. linearis Fl. Rocky Mts. 818, 1066. 1917 [1918] **Utah:** Green River, 19 Aug 1887, *Tracy & Evans 716* (HT: NY!).

SAXIFRAGACEAE

Chrysosplenium iowensis Rydb. *in* Britton, Man. fl. n. states 483. 1901. **Iowa:** Decorah, 5 May 1889, *Holway s.n.* (HT: NY!).

Conimitella Rydb. in Small & Rydb., NAF **22:** 96. 1905. T: *Heuchera williamsii* D. C. Eat.

Elmera Rydb. *in* Small & Rydb., NAF **22:** 97. 1905. T: *Heuchera racemosa* S. Wats.

Heuchera abramsii Rydb. *in* Small &Rydb., NAF **22:** 109. 1905. **California. San Bernardino Co.:** Mt. San Antonio, Jul 1901, *Abrams 1924* (HT: NY!; IT: NY!).

H. aceroides Rydb. ex Small, Fl. S.E. U.S. 503, 1331. 1903. **North Carolina:** Asheville, May 1886, *Hogg s.n.* (HT: NY!).

H. alba Torreya **26:** 31. 1926. **West Virginia. Pendleton Co.:** Snowy Mt., 17 Jun 1925, *Rydberg 9078* [*9026* in publciation] (HT: NY!; IT: NY!).

H. arkansana Rydb. *in* Small & Rydb., NAF **22:** 101. 1905. **Arkansas. Benton Co.:** near Springdale, 16 Aug 1895, *Blankinship s.n.* (HT: NY!).

H. ciliata Mem. New York Bot. Gard. **1:** 196. 1900. **Montana:** Mill Creek, Aug 1887, *Tweedy 259* (HT: YU!; CT: NY!).

H. columbiana Rydb. *in* Small & Rydb., NAF **22:** 116. 1905. **Washington:** Swan Lake, 20 Jul 1897, *Winston s.n.* (HT: NY!).

H. crinita Rydb. *in* Britton, Man. fl. n. states 482. 1901. **Kentucky:** *Short s.n.* (HT: NY!).

H. diversifolia Rydb. *in* Small & Rydb., NAF **22:** 102. 1905. **Washington. Chehalis Co.:** 24 Jul 1897, *Lamb 1305* (HT: NY!).

H. flabellifolia Rydb. in Small & Rydb., NAF **22:** 115. 1905. **Montana. Gallatin Co.:** Bridger Mts., 18 Jun 1897, *Rydberg & Bessey 4292* (HT: NY!; IT: NY!).

H. flavescens Rydb. *in* Small & Rydb., NAF **22:** 114. 1905. **New Mexico:** Santa Fe Canyon, 9 miles east of Santa Fe, 11 Jun 1897, *Heller & Heller 3693* (HT: NY!; IT: NDG, NEB, US!).

H. glaberrima Rydb. *in* Small & Rydb., NAF **22:** 103. 1905. **Oregon:** *Howell s.n.* (HT: NY!).

H. gracilis Rydb. *in* Small & Rydb., NAF **22:** 114. 1905. **Idaho:** Mt. Idaho, 28 Jun 1895, *Henderson 3159* (HT: US!; CT: NY!).

H. grossulariifolia Mem. New York Bot. Gard. **1:** 196. 1900. **Montana:** Pony, 6 Jul 1897, *Rydberg & Bessey 4288* (HT: NY!—designated by Rydberg in herb.; IT: CAN, MONT, NEB).

H. halstedii Rydb. *in* Small & Rydb., NAF **22:** 104. 1905. **MEXICO:** Hacienda de Regla, *Halsted s.n.* (HT: NY!).

H. hemisphaerica Rydb. *in* Small & Rydb., NAF **22:** 112. 1905. **California:** Santa Lucia Mts., 1886, *Brandegee s.n.* (HT: GH!).

H. hirsuta Rydb. *in* Small & Rydb., NAF **22:** 109. 1905. **California. San Bernardino Co.:** Snow Canyon, 20 Jun 1901, *Parish 5062* (HT: NY!).

H. lancipetala Rydb. *in* Britton, Man. fl. n. states 482. 1901. **Kentucky:** 1840, *Short s.n.* (HT: NY!).

H. lloydii Rydb. *in* Small & Rydb., NAF **22:** 113. 1905. **[Oregon]:** Cascade Mts., 45th Parallel, 1895, *Lloyd s.n.* (HT: NY!).

H. longiflora Rydb. *in* Britton, Man. fl. n. states 482. 1901. KENTUCKY: Hills bordering Mudlick Bath, *Short s.n.* (HT: NY!).

H. mexicana Schaffner ex Rydb. *in* Small & Rydb., NAF **22:** 108. 1905. **MEXICO.** **SAN LUIS POTOSÍ:** near Morelos, Aug 1876, *Schaffner 450* (HT: NY!; IT: US! dated 1879).

H. nuttallii Rydb. *in* Small & Rydb., NAF **22:** 103. 1905. Rocky Mts., Columbia River, *Nuttall s.n.* (HT: NY!).

H. parishii Rydb. *in* Small & Rydb., NAF **22:** 109. 1905. CALIFORNIA: San Bernardino Mts., Mill Creek, 3 Jul 1892, *Parish 2512* (HT: NY!).

H. pilosella Rydb. *in* Small & Rydb., NAF **22:** 112. 1905. CALIFORNIA: San Luis Obispo, 1876, *Palmer 124* (HT: NY!; IT: US! labeled Central California).

H. pringlei Rydb. *in* Small & Rydb., NAF **22:** 111. 1905. CALIFORNIA. **Siskiyou Co.:** Castle Lake, 14 Aug 1881, *Pringle s.n.* (HT: NY!).

H. pulchra Rydb. *in* Small & Rydb., NAF **22:** 117. 1905. ARIZONA: Sabenio Canyon, 30 Mar 1901, *Griffiths 2565* (HT: NY!; IT: NY! dated 27 Mar).

H. reglensis Rydb. *in* Small & Rydb., NAF **22:** 108. 1905. **MEXICO.** Regla, *Halsted s.n.* (HT: NY!).

H. roseola Rydb. *in* Britton, Man. fl. n. states 481. 1901. PENNSYLVANIA: York Furnace, 29 May 1892, *Brinton s.n.* (HT: NY!).

H. scabra Rydb. *in* Small & Rydb., NAF **22:** 106. 1905. NORTH CAROLINA. **Madison Co.:** Hot Springs, 1899, *Churchill s.n.* (HT: NY!).

H. sitgreavesii Rydb. *in* Small & Rydb,. NAF **22:** 110. 1905. ARIZONA: Camp 19 between San Francisco and Bill Williams Mts., 21 Oct 1851, *Woodhouse s.n.* (HT: NY!).

H. suksdorfii Rydb. *in* Small & Rydb., NAF **22:** 116. 1905. WASHINGTON: Falcon Valley, 28 Jun 1882, *Suksdorf s.n.* (HT: NY!).

H. townsendii Rydb. *in* Small & Rydb., NAF **22:** 117. 1905. **MEXICO.** CHIHUAHUA: Sierra Madre near Colonia García, 30 May 1899, *Townsend & Barber 3* (HT: NY!; IT: US!).

H. utahensis Rydb. *in* Small & Rydb., NAF **22:** 114. 1905. UTAH: City Creek Canyon, 17 Jul 1880, *Jones 1458* [*9458* in publication] (HT: NY!; IT: BRY, CAS, NY! 2 sheets, POM, RM, UC, US!, UT!, UTC!).

H. wootonii Rydb. *in* Small & Rydb., NAF **22:** 113. 1905. NEW MEXICO. **Lincoln Co.:** White Mts., 5 Aug 1897, *Wooton 283* (HT: NY!; IT: US 2 sheets).

Lithophragma australis Rydb. *in* Small & Rydb., NAF **22:** 86. 1905. [ARIZONA]: Cedar Creek, Camp 79, 24 Apr 1858, *Newberry s.n.* (HT: NY!; IT: US! no collector on label).

L. breviloba Rydb. *in* Small & Rydb., NAF **22:** 86. 1905. CALIFORNIA. **Sierra Co.:** 1874, *Lemmon 647* (HT: GH!).

L. bulbifera Rydb. *in* Small & Rydb., NAF **22:** 86. 1905. WYOMING. **Carbon Co.:** Battle, 20 Jun 1901, *Tweedy 4411* (HT: NY!; IT: US!, YU!).

L. catalinae Rydb. *in* Small & Rydb., NAF **22:** 88. 1905. CALIFORNIA: Santa Catalina Island, Avalon, Mar 1901, *Trask s.n.* (HT: NY!).

L. intermedia Rydb. *in* Small & Rydb., NAF **22:** 88. 1905. CALIFORNIA. **Los Angeles Co.:** 7 Apr 1892, *Hasse s.n.* (HT: NY!).

L. laciniata Eastw. ex Rydb. *in* Small & Rydb., NAF **22:** 87. 1905. CALIFORNIA. **Butte Co.:** Marysville, 1893, *Blankinship s.n.* (HT: CAS!).

L. trifida Eastw. ex Rydb. *in* Small & Rydb., NAF **22:** 89. 1905. CALIFORNIA. **San Diego Co.:** Near Sweetwater Dam, 10 Mar 1891, *Eastwood s.n.* (HT: CAS!).

L. triloba Rydb. in Small & Rydb., NAF **22:** 87. 1905. CALIFORNIA. **Monterey Co.:** Santa Lucia Mts., Feb 1898, *Plaskett 27* (HT: NY!; IT: US!).

Mitella intermedia Bruhin ex Rydb. *in* Small & Rydb., NAF **22:** 92. 1905. WISCONSIN. **Manitowoc.:** Near Lake Michigan, one mile north of Centerville, 1876, *Bruhin s.n.* (HT: US!).

M. oppositifolia Rydb. *in* Small & Rydb., NAF **22:** 91. 1905. MASSACHUSETTS. **[Hampshire Co.]:** South Hadley, May 1904, *Kinney s.n.* (HT: NY!).

M. violacea BTBC **24:** 248. 1897. MONTANA: Madison Range, Spanish Basin, 11 Jul 1896, *Flodman 527* (HT: NY! on 2 sheets).

Ozomelis pacifica Rydb. *in* Small & Rydb., NAF **22:** 95. 1905. WASHINGTON. **Challam Co.:** Olympic Mts., Aug 1900, *Elmer 2778* (HT: NY!).

Parnassia floridana NAF **22:** 80. 1905. **FLORIDA. [Gulf Co.]:** Appalachicola, *Chapman s.n.* (HT: NY!).

P. intermedia NAF **22:** 78. 1905. NEVADA: east Humboldt [Ruby] Mts., Sep 1868, *Watson 370* (HT: NY!; IT: GH, US, YU!).

P. montanensis Fernald & Rydb. *in* Rydb., NAF **22:** 79. 1905. MONTANA: Sixteen Mile Creek, 12 Jul 1883, *Scribner 54* (HT: GH; IT: NY!).

P. rivularis Osterhout ex Rydb., NAF **22:** 78. 1905. COLORADO: North Park, 25 Jul 1898, *Osterhout 1778* (HT: RM!; IT: NY!).

Pectiantia latiflora Rydb. *in* Small & Rydb., NAF **22:** 93. 1905. WASHINGTON: Palace Camp, 1883, *Willis s.n.* (HT: NY!).

Saxifraga rydbergii Small ex Rydb., Mem. New York Bot. Gard. **1:** 194. 1900. MONTANA: Yellowstone Park, Electric Peak, 18 Aug 1897, *Rydberg & Bessey 4268* (HT: NY!).

Sullivantia halmicola A. Nels. ex Rydb. *in* Small & Rydb., NAF **22:** 122. 1905. WYOMING: Casper Mt., Hat Six Creek, 7 Aug 1898, *E. Nelson 5032* (HT: NY!).

Therophon cincinnatum Rosendahl & Rydb. *in* Small & Rydb., NAF **22:** 124. 1905. CALIFORNIA: Santa Cruz, 18 Jul 1882, *Pringle s.n.* (HT: NY!; IT: CAN, NY!, US!). Gornall & Bohm, J. Linn. Soc., Bot. **90:** 41. 1985, considered the CAN specimen to be the holotype but this arbitrary selection has no justification.

T. [*Therofon* in publication] *heucheraeforme* BTBC **24:** 247. 1897. MONTANA: Bridger Mts., 28 Jul 1896, *Flodman 514* (LT: NY!—Rydberg, NAF **22:** 126. 1905). Gornall & Bohm, J. Linn. Soc., Bot. **90:** 41, 1985, selected *Rydberg 2677* (MIN) as a lectotype but I consider Rydberg's 1905 reference to a type locality as a lectotypification.

T. turbinatum Rydb. *in* Small & Rydb., NAF **22:** 124. 1905. ALABAMA: Mountains of northern Alabama,

Sep 1840, *Buckley s.n.* (HT: NY!; IT: NY! labeled just Alabama).

T. vancouverense Rydb. *in* Small & Rydb., NAF **22:** 125. 1905. **CANADA. [British Columbia]:** Vancouver Island, 1 Aug 1874, *Edwards s.n.* (HT: NY!).

SCROPHULARIACEAE

Adenostegia ciliosa BTBC **34:** 35. 1907. **Wyoming:** Spread Creek, Jul 1897, *Tweedy 245* [*545* in publication] (HT: NY!).

Besseya BTBC **30:** 279. 1903. T: *Synthyris alpina* A. Gray.

Castilleja ampliflora BTBC **34:** 39. 1907. **Montana:** Divide between McDonald and Camas Lakes, 29 Jul 1901, *Vreeland 995* (HT: NY!; IT: NY!).

C. arcuata BTBC **34:** 35. 1907. **Utah:** South end of Fish Lake, 10 Aug 1905, *Rydberg & Carlton 7508* (HT: NY!; IT: RENO!, RM, US!).

C. brunnescens BTBC **31:** 643. 1904 [1905]. **Colorado:** Cameron Pass, 15 Jul 1896, *Baker s.n.* (HT: NY!).

C. crista-galli Mem. New York Bot. Gard. **1:** 355. 1900. **Montana:** Bridger Mts., 17 Jun 1897, *Rydberg & Bessey 4950* (HT: NY!; IT: NY!, US!).

C. gracillima BTBC **34:** 39. 1907. **Wyoming:** Yellowstone Park, Lower Geyser Basin, 7 Aug 1897, *Rydberg & Bessey 4964* (HT: NY!; IT: NY!, US!).

C. humilis BTBC **34:** 37. 1907. **Wyoming. Albany Co.:** Medicine Bow Mts., 2 Aug 1900, *Nelson 7919* (HT: NY!).

C. lanceifolia Mem. New York Bot. Gard. **1:** 357. 1900. **Montana. Gallatin Co.:** Spanish Basin, 1 Jul 1897, *Rydberg & Bessey 4954* (HT: NY!; IT: CAN, MONT, NEB, NY!, US!).

C. leonardii BTBC **34:** 36. 1907. **Utah:** Head of American Fork Canyon, 21 Jul 1885, *Leonard 151* in part (HT: NY!).

C. linearis BTBC **28:** 25. 1901. **Colorado:** Near West Spanish Peak, 6 Jul 1900, *Rydberg & Vreeland 5619* (HT: NY!).

C. luteovirens BTBC **28:** 26. 1901. **Colorado:** Sangre de Cristo Creek, 2 Jul 1900, *Rydberg & Vreeland 5616* (HT: NY!).

C. magna BTBC **34:** 36. 1907. **CANADA. British Columbia:** above Carbonate, 8 Jul 1904, *Shaw 205* (HT: NY!).

C. obtusiloba BTBC **31:** 644. 1904 [1905]. **Colorado. Delta Co.:** Leroux Park, 19 Jul 1892, *Cowan s.n.* (HT: NY!).

C. parvula BTBC **34:** 40. 1907. **Utah:** Mountains north of Bullion Creek, near Marysvale, 23 Jul 1905, *Rydberg & Carlton 7158* (HT: NY!; IT: US!).

C. pecten BTBC **34:** 41. 1907. **Idaho:** Beaver Canyon, 28 Jun 1895, *Shear 3041* (HT: NY!).

C. puberula BTBC **31:** 644. 1904 [1905]. **Colorado:** 1872, *Parry s.n.* (HT: NY!).

C. pulchella BTBC **34:** 40. 1907. **Montana:** Mountains near Indian Creek, 21 Jul 1897, *Rydberg & Bessey 4967* (HT: NY!; IT: MONT, NEB).

C. purpurascens BTBC **34:** 38. 1907. **CANADA. British Columbia:** Flood plains of Kicking Horse, 24 Jun 1904, *Petersen 11* (HT: NY!).

C. rhexifolia Mem. New York Bot. Gard. **1:** 356. 1900. **Montana:** Cedar Mt., 16 Jul 1897, *Rydberg & Bessey 4951* (HT: NY!; IT: NY!, US!).

C. stricta Mem. New York Bot. Gard. **1:** 354. 1900. **Nevada:** Ruby Valley, Aug 1868, *Watson 809* (HT: NY!; IT: US!, YU!).

C. subcinerea BTBC **40:** 484. 1913. **Idaho:** Beaver Canyon, 28 Jun 1895, *Shear 3041* (HT: NY!).

C. sulphrea Mem. New York Bot. Gard. **1:** 359. 1900. **Montana:** Electric Peak, 20 Aug 1897, *Rydberg & Bessey 4966* (HT: NY!).

C. trinervis BTBC **28:** 26. 1901. **Colorado:** Headwaters of Sangre de Cristo Creek, 22 Jun 1900, *Rydberg & Vreeland 5620* (HT: NY!).

C. tweedyi Mem. New York Bot.Gard. **1:** 358. 1900. **Montana:** Jack Creek, 14 Jul 1897, *Rydberg & Bessey 4962* (HT: NY!; IT: CAN, MONT, NY!).

C. variabilis BTBC **34:** 37. 1907. **Utah:** Big Cottonwood Canyon, below Silver Lake, 8 Jul 1905, *Rydberg 6773* (HT: NY!; IT: RM).

C. villosa Mem. New York Bot. Gard. **1:** 361. 1900. **Montana:** Valley of Blackfoot River, 15 Jul 1883, *Canby 261* (HT: NY!).

C. viscida BTBC **34:** 38. 1907. **Utah:** Mountains near the headwaters of Big Cottonwood Creek, 3 Jul 1905, *Rydberg & Carlton 6593* (HT: NY!).

C. vreelandii BTBC **34:** 38. 1907. **Montana:** Divide between McDonald and Camas lakes, 29 Jul 1901, *Vreeland 1000* (HT: NY!).

C. wyomingensis BTBC **28:** 502. 1901. **Wyoming. Sheridan Co.:** Big Horn Mts., Jul 1899, *Tweedy 2341* (HT: NY!; IT: YU!).

Elephantella Mem. New York Bot. Gard. **1:** 362. 1900. T. *Pedicularis groenlandica* Retz.

Mimulus eastwoodiae BTBC **40:** 483. 1913. **Utah:** Along San Juan River, near Bluffs, 25–29 Aug 1911, *Rydberg 9883* (HT: NY! on 2 sheets; IT: RM, US!, UT!).

M. gratioloides BTBC **28:** 27. 1901. **Colorado:** Butte 5 miles southwest of La Veta, 22 May 1900, *Rydberg & Vreeland 5660* (HT: NY!: IT: NEB).

M. puberulus E. L. Greene ex Rydb., Fl. Colorado 311. 1906. **Colorado:** Pagosa Springs, 27 Jul 1899, *Baker 587* (HT: NY!; IT: US!).

Pedicularis ctenophora BTBC **24:** 293. 1897. **Montana:** Lima, 29 Jul 1895, *Rydberg 2789* (HT: NY!).

P. cystopteridifolia Mem. New York Bot. Gard. **1:** 365. 1900. **Montana:** Cedar Mt., 16 Jul 1897, *Rydberg & Bessey 4983* (HT: NY!; IT: CAN, MONT, NY!, US!).

P. hallii Mem. New York Bot. Gard. **1:** 364. 1900. **Wyoming:** Yellowstone Park, Jun 1888, *C. H. Hall s.n.* (HT: NY!).

P. lunata BTBC **28:** 27. 1901. **Wyoming:** Big Horn Mts., Aug 1899, *Tweedy 2317* (HT: NY!).

P. montanensis BTBC **24:** 292. 1897. **Montana:** Little Belt Mts., 9 miles from Barker, 18 Aug 1896, *Flodman 796* (HT: NY!).

P. siifolia BTBC **34:** 35. 1907. **Montana:** Grant Creek, 7 Jun 1897, *Elrod 97* (HT: NY!).

Penstemon aridus Mem. New York Bot. Gard. **1:** 348. 1900. **Montana. Gallatin Co.:** Spanish Basin, 23 Jun 1897, *Rydberg & Bessey 4920* (HT: NY!; IT: CAN, MONT, NEB 2 sheets, US!).

P. auricomus A. Nels. ex Rydb., BTBC **36:** 688. 1909. **Wyoming. Sweetwater Co.:** Green River, 31 May 1897, *Nelson 3052* (HT: NY!).

P. coccinatus BTBC **36:** 691. 1909. **Arizona:** Grand Canyon of the Colorado, 28 Jun 1898, *MacDougal 173* (HT: NY!).

P. cyathophorus BTBC **31:** 643. 1904 [1905]. **Colorado:** Pearl, 9 Jul 1901, *Tweedy 4307* (HT: NY!; IT: US!).

P. diphyllus Mem. New York Bot. Gard. **1:** 349. 1900. **Montana:** Mullen Pass, Aug 1860, *Cooper s.n.* (ST: NY!); **Silver Bow Co.:** Soap Gulch, Jul 1888, *Tweedy 72* (ST: YU!).

P. erosus BTBC **28:** 28. 1901. **Colorado:** Indian Creek Pass, 1900, *Vreeland 615* (HT: NY not found).

P. leonardii BTBC **40:** 483. 1913. **Utah:** Diehl's Grove, Wasatch Mts., 1 Aug 1884, *Leonard s.n.* (HT: NY!).

P. leptophyllus Fl. Rocky Mts. 773, 1066. 1917 [1918]. **Idaho. Washington Co.:** Rush Creek, 10 Jul 1899, *Jones 6493* (HT: NY!).

P. oreophilus BTBC **31:** 642. 1904 [1905]. **Colorado:** Eldora to Baltimore, 20 Jun–10 Jul 1903, *Tweedy 5711* (HT: NY!).

P. pseudohumilis Mem. New York Bot. Gard. **1:** 347. 1900. **Idaho:** Mt. Chauvet, 29 Jul 1897, *Rydberg & Bessey 4915* (HT: NY!; IT: MONT, US!).

P. pseudoprocerus Mem. New York Bot. Gard. **1:** 346. 1900. **Montana:** Bridger Mt., 11 Jun 1897, *Rydberg & Bessey 4919* (HT: NY!; IT: CAN, MONT, NEB 2 sheets, NY!, US! all dated 12 Jun).

P. puberulentus Fl. Rocky Mts. 774, 1066. 1917 [1918]. **Idaho:** Long Valley, 6 Jul 1895, *Henderson 3169* (HT: US!; IT: NY!).

P. saliens Mem. New York Bot. Gard. **1:** 344. 1900. **Montana:** Columbia Falls, 1894, *Kennedy 53* (HT: NY!; IT: MONT).

P. strictiformis BTBC **31:** 642. 1904 [1905]. **Colorado:** Mancos, 23 Jun 1898, *Baker, Earle & Tracy 76* (HT: NY!; IT: NEB, US!).

P. suffrutescens BTBC **28:** 503. 1901. **Colorado:** Ridgway, 20 Jun 1895, *Tweedy 170* (HT: NY!; IT: YU!).

P. virens Pennell ex Rydb., Fl. Rocky Mts. 773, 1066. 1917 [1918]. **Colorado: Jefferson Co.:** West of Morrison, 13 Jun 1915, *Pennell 5821* (HT: NY!).

P. wilcoxii BTBC **28:** 28. 1901. **Montana:** Kalispel, 2 Jun 1900, *Wilcox 370* (HT: US; IT: NY! dated just 1900).

Penstemonopsis Fl. Rocky Mts. 777, 1066. 1917 [1918]. T: *Penstemon tweedyi* Canby & Rose.

Scrophularia neglecta Rydb. ex Small, Fl. S.E. U.S. 1058, 1337. 1903. **Kansas. Riley Co.:** 1896, *Norton 779* (HT: NY!; IT: US!).

S. nodosa var. *occidentalis* Contr. U.S. Natl. Herb. **3:** 517. 1896. **South Dakota:** Rapid City, 25 Jul 1892, *Rydberg 914* (HT: US!; IT: NY!).

S. serrata BTBC **36:** 688. 1909. **Idaho:** Granite, Jul 1887, *Sandberg s.n.* (HT: NY!).

Synthyris dissecta BTBC **36:** 691. 1909. **Montana. Gallatin Co.:** Near Bozeman, 11 Jun 1900, *Chesnut & Jones 199* (HT: NY!).

Veronica americana var. *crassula* Mem. New York Bot. Gard. **1:** 353. 1900. **Montana.** Little Belt Pass, 10 Aug 1896, *Flodman 778* (HT: NY!).

SELAGINELLACEAE

Selaginella densa Mem. New York Bot. Gard. **1:** 7. 1900. **Montana:** Little Rocky Mts., Sep 1889, *Havard s.n.* (LT: NY!—Tryon, Ann. Missouri Bot. Gard. **42:** 66. 1955; CT: US!).

SOLANACEAE

Chamaesaracha crenata Mem. Torrey Bot. Club **4:** 368. 1896. **MEXICO. Coahuila:** Parras, Feb–Oct 1880, *Palmer 923* (HT: US!; IT: NY! dated 8–28 Jun).

C. villosa Mem. Torrey Bot. Club **4:** 368.1896. **MEXICO. Coahuila & Nuevo León:** Feb–Oct 1880, *Palmer 924* (HT: US!).

Margaranthus purpurascens Mem. Torrey Bot. Club **4:** 317. 1896. **New Mexico:** Mogollon Mts., 6 Sep 1881, *Rusby 307* (HT: US!).

Physalis ciliosa Mem. Torrey Bot. Club **4:** 346. 1896. **Florida:** *Chapman s.n.* (LT: GH!—Waterfall, Rhodora **60:** 139. 1958; ILT: GH!, NY!, OKL).

P. comata BTBC **22:** 306. 1895. **Nebraska:** South side of Scotts Bluff, 20 Jul 1891, *Rydberg 269* (HT: NY! on 2 sheets; IT: NEB).

P. floridana Rydb. *in* Small, Fl. S.E. U.S. 983, 1337. 1903. **Florida. Walton Co.:** 1885, *Curtiss s.n.* (HT: NY!; IT: NY! 3 sheets).

P. hastata Mem. Torrey Bot. Club **4:** 363. 1896. **MEXICO. Lower California:** San José del Cabo, 22 Sep 1890, *Brandegee 422* (HT: US!).

P. heterophylla var. *umbrosa* Contr. U.S. Natl. Herb. **3:** 172. 1895. **Nebraska:** Middle Loup River near Thedford, 21 Jun 1893, *Rydberg 1398* (HT: US!; IT: NY! 2 sheets, US! dated 22 Aug).

P. macrophysa BTBC **22:** 308. 1895. **Texas. [Kerr Co.]:** Kerrville, 17 May 1894, *Heller 1756* (LT: NY!—Waterfall, Rhodora **60:** 153. 1958; ILT: NY!, UC).

P. neomexicana Mem. Torrey Bot. Club **4:** 325. 1895. **New Mexico:** 1847, *Fendler 678* (LT: GH!—Waterfall, Rhodora **60:** 169. 1958; ILT: GH! 2 sheets).

P. pendula Rydb. *in* Small, Fl. S.E. U.S. 983, 1337. 1903. **[Arkansas]:** Central and southern Arkansas, *Harvey 65* (LT: UARK—Waterfall, Rhodora **60:** 163. 1958).

P. rotundata Mem. Torrey Bot. Club **4:** 352. 1896. **South Dakota:** Spring Creek, 19 Aug 1891, *Williams s.n.* (HT: NY!).

P. sinuata Rydb. *in* Small, Fl. S.E. U.S. 986, 1337. 1903. **Florida:** *Chapman s.n.* (HT: NY!).

P. subulata BTBC **22:** 306. 1895. **MEXICO Chihuahua:** Guerrero, 8 Sep 1887, *Pringle 1344* (HT: NY! on 2 sheets; IT: US!).

P. texana Mem. Torrey Bot. Club **4:** 339. 1896. **Texas. Nueces Co.:** Corpus Christi, 27 May 1894, *Heller 1507* (HT: NY!; IT: ARIZ, GH, NY! 2 sheets, PH, UC, US! dated 23–30 May).

P. versicolor BTBC **22:** 307. 1895. **MEXICO. [Sonora]:** Guaymas, Nov 1887, *Palmer 622* (LT: NY!—Waterfall, Rhodora **60:** 160. 1958; ILT: GH!).

P. versicolor var. *microphylla* BTBC **22:** 307. 1895. **MEXICO. [Sonora]:** Guaymas, Jul 1887, *Palmer 94* (HT: NY!; IT: GH!, US! dated 1887).

P. virginiana var. *intermedia* Mem. Torey Bot. Club **4:** 345. 1896. ALABAMA: Central Alabama, *Buckley 1* (HT: NY!).

Solanum agrimonifolium BTBC **51:** 154. 1924 **MEXICO.** CHIAPAS: Cerro del Boquerón, Sep 1913, *Purpus 6977* (HT: US!; IT: NY!).

S. canescens BTBC **51:** 174. 1924. **PANAMA.** Vicinity of Cana, 17 Apr–8 Jun 1908, *Williams 844* (HT: NY! on 3 sheets).

S. ervendbergii BTBC **51:** 153. 1924. **MEXICO. [VERACRUZ]:** Huasteca, Wartenberg, near Tantoyuca, 1858, *Ervendberg 175* (HT: GH!).

S. inscendens BTBC **51:** 175. 1924. **MEXICO. [MÉXICO]:** Ixtaccíhuatl, Jan 1906, *Purpus 1733* (HT: NY!; IT: US).

S. interius BTBC **31:** 641. 1904 [1905]. **NEBRASKA. Hooker Co.:** On Middle Loup River, near Mullen, 20 Jul 1893, *Rydberg 1385* (HT: NY!; IT: US!).

S. nicaraguense BTBC **51:** 171. 1924. **NICARAGUA.** 1868, *Flint 8* (HT: US!).

S. papita BTBC **51:** 148. 1924. **MEXICO. DURANGO:** Otinapa, 25 Jul–5 Aug 1906, *Palmer 392* (HT: US!; IT: NY!).

S. polytrichon BTBC **51:** 150. 1924. **MEXICO.** Vicinity of San Luis Postosí, 22° N. Lat., 1878, *Parry & Palmer 632* (HT: US!).

S. sambucinum BTBC **51:** 169. 1924. **MEXICO. QUERÉTARO:** Near San Juan del Río, 18 Aug 1905, *Rose, Painter & Rose 9586* (HT: US!).

S. subvelutinum BTBC **51:** 175. 1924. **MEXICO. DURANGO:** San Ramón, Apr or May 1906, *Palmer 183* (HT: US not found).

S. wightianum BTBC **51:** 149. 1924. Type grown from tubers, collected near **MEXICO. GUANAJUATO:** Acámbaro, Oct 1904, *Lozano s.n.* (HT: GH; IT: US! dated just 1904).

SPARGANIACEAE

Sparganium chlorocarpum NAF **17:** 8. 1909. **IOWA. Emmet Co.:** Sep 1898, *Cratty s.n.* (HT: NY!).

S. williamsii NAF **17:** 10. 1909. **CANADA. YUKON TERRITORY:** Klondike Bottom, 9 & 23 Jul 1899, *Williams s.n.* (HT: NY!; IT: US!).

ULMACEAE

Celtis rugosa BTBC **39:** 304. 1912. **COLORADO:** Golden, 29 Aug 1895, *Rydberg s.n.* (HT: NY! on 2 sheets).

C. villosula Fl. Rocky Mts. ed. 2, 1116. 1922 [1923]. **UTAH:** 1895, *Jones s.n.* (HT: NY!).

URTICACEAE

Parietaria obtusa Rydb. ex Small, Fl. S.E. U.S. 359, 1329. 1903. **UTAH:** Southern Utah, 1874, *Palmer 226* (HT: NY!).

P. occidentalis BTBC **39:** 306. 1912. **IDAHO. Nez Perces Co.:** Clearwater River, 14 May 1892, *Sandberg, MacDougal & Heller 176* (HT: NY!).

Urtica cardiophylla BTBC **24:** 191. 1897. **MONTANA:** Near Castle, 1 Aug 1896, *Flodman 370* (HT: NY!).

U. strigosissima BTBC **39:** 305. 1912. **IDAHO. Nez Perces Co.:** 29 Jul 1896, *Heller & Heller 3475* (HT: NY!; IT: US!).

U. viridis BTBC **39:** 305. 1912 **MONTANA:** Emigrant Gulch, 23 Aug 1897, *Rydberg & Bessey 3935* (HT: NY!; IT: CAN, MONT, NEB).

VALERIANACEAE

Valeriana acutiloba BTBC **28:** 24. 1901. **COLORADO:** Sangre de Cristo Range, near Gray Back Mining Camps and Placer Gulch, 25–27 Jun 1900, *Rydberg & Vreeland 5576* (HT: NY!).

V. ovata BTBC **31:** 645. 1904 [1905]. **COLORADO:** Cameron's Cone, 26 Jun 1901, *Clements & Clements 241* (HT: NY!; IT: NEB).

V. puberulenta BTBC **36:** 697. 1909. **UTAH:** Mountains north of Bullion Creek, near Marysvale, 23 Jul 1905, *Rydberg & Carlton 7065* (HT: NY! on 2 sheets; IT: BRY, US!).

V. pubicarpa BTBC **36:** 697. 1909. **UTAH:** Mt. Nebo, 15 Aug 1905, *Rydberg & Carlton 7717* (HT: NY!; IT: RM, US!).

V. trachycarpa BTBC **31:** 645. 1904 [1905]. **COLORADO:** Red Mt., south of Ouray, 11 Sep 1901, *Underwood & Selby 352* (HT: NY!).

VERBENACEAE

Verbena ambrosifolia Rydb. ex Small, Fl. S.E. U.S. 1011, 1337. 1903. **COLORADO:** Rocky Ford, 8 Jun 1900, *Osterhout s.n.* (HT: NY!).

V. pumila Rydb. ex Small, Fl. S.E. U.S. 1010, 1337. 1903. **[OKLAHOMA]:** Indian Territory, Catoosa, 8 May 1895, *Bush 1275* (HT: NY!).

VIOLACEAE

Viola montanensis Mem. New York Bot. Gard. **1:** 263. 1900. **MONTANA:** Jack Creek Canyon, 15 Jul 1897, *Rydberg & Bessey 4532* (ST: MONT, NY! 2 sheets); Bridger Mts., 12 Jun 1897, *Rydberg & Bessey 4531* (ST: CAN, NY!, NEB).

V. monticola Mem. New York Bot. Gard. **1:** 264. 1900. **MONTANA:** Cedar Mt., 16 Jul 1897, *Rydberg & Bessey 4529* (ST: CAN, NY!); Electric Peak, 20 Aug 1897, *Rydberg & Bessey 4527*; Bozeman, 1885, *Tweedy 456;* 1892, *Shaw s.n.*; Preist's Pass, 1892, *Kelsey s.n.* **WYOMING:** Yellowstone National Park, Lone Star Geyser, 7 Aug 1897, *Rydberg & Bessey 4526* (ST: NY!); Yellowstone National Park, Mammoth Hot Springs, 1889, *Dewart s.n.*; 1884, *Tweedy 103.*

V. odontophora Mem. New York Bot. Gard. **1:** 264. 1900. **MONTANA:** Crafton, 2 Jun 1892, *Williams 114* (HT?: MONT).

VISCACEAE

Razoumofskya cyanocarpa A. Nels. ex Rydb., Fl. Colorado 100–101. 1906. **Wyoming:** Ferris Mts., 24 Jul 1898, *Nelson 4959* (HT: NY!).

ZYGOPHYLLACEAE

Fagonia pachyacantha Rydb. *in* Vail & Rydb., NAF **25:** 105. 1910. **MEXICO. Lower California:** *Diquet s.n.* (HT: NY!).

Guaiacum guatemalense Planchon ex Rydb. *in* Vail & Rydb., NAF **25:** 106. 1910. **GUATEMALA. Zacapa:** *Skinner s.n.* (HT: NY!).

G. planchoni A. Gray ex Rydb. *in* Vail & Rydb., NAF **25:** 107. 1910. **MEXICO. Oaxaca:** Between Tehuantepec and Pacific Ocean, *Andrieux 475* (HT: NY!; IT: K).

Kallstroemia canescens Rydb. *in* Vail & Rydb., NAF **25:** 113. 1910. **MEXICO. Sinaloa:** Rosario, 7 Jul 1897, *Rose 1547* (HT: US!; IT: NY!).

K. caribea Rydb. *in* Vail & Rydb., NAF **25:** 111. 1910. **WEST INDIES. Montserrat:** Coconut Hill, 5 Feb 1907, *Shafer 388* (HT: NY!; IT: US!).

K. curta Rydb. *in* Boldingh, Fl. voor de Nederlandsch West-Indische eilanden 230. 1913. **CURAÇAO:** St. Jores Bay, 20–27 Mar 1913, *Britton & Shafer 3096* (HT: NY!—designated by Rydberg in herb.; IT: US!).

K. glabrata Rydb. *in* Vail & Rydb. NAF **25:** 112. 1910. **MEXICO. [Guerrero]:** Acapulco, Oct 1894–Mar 1895, *Palmer 131* (HT: US!).

K. incana Rydb. *in* Britton, BTBC **51:** 3. 1924. **[DOMINICAN REPUBLIC]. Barahona:** Jul 1910, *Fuertes 418* (HT: NY!; IT: US!).

K. intermedia Rydb. *in* Vail & Rydb., NAF **25:** 113. 1910. **Texas. Bexar Co.:** 1904, *Jermy s.n.* (HT: NY!).

K. longipes Rydb. *in* Vail & Rydb., NAF **25:** 114. 1910. **MEXICO. Sinaloa:** Rosario, 22 Jul 1897, *Rose 1829* (HT: NY!; IT: US!).

K. rosei Rydb. *in* Vail & Rydb., NAF **25:** 113. 1910. **MEXICO. Morelos:** Yautepec, 27 Aug 1903, *Rose & Painter 6562* (HT: US!; IT: NY!).